Growth and Ecology of Fish Populations

Growth and Ecology of Fish Populations

A. H. WEATHERLEY
Zoology Department, Australian National University, Canberra, Australia

1972

ACADEMIC PRESS · London · New York

ACADEMIC PRESS INC. (LONDON) LTD.
24/28 Oval Road,
London NW1

United States Edition published by
ACADEMIC PRESS INC.
111 Fifth Avenue
New York, New York 10003

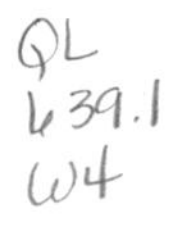

Library of Congress Card Number: 79-175833
ISBN: 0 12 739050 2

PRINTED IN GREAT BRITAIN BY
The Whitefriars Press Ltd., London and Tonbridge

Preface

Few other groups of highly evolved animals enjoy greater mastery of their environment than fish—a mastery characterized by a great proliferation of functional and structural adaptations to various habitats and conditions. Fish taxonomy and systematics are thriving—though often contentious—fields of biological research, and animal physiologists have been interested in fish as experimental animals for at least 200 years. Moreover, the great economic significance of many of the world's fisheries—marine and inland—has engendered a great deal of study of many aspects of fish population dynamics. Yet there have been fewer general works on their ecology than might have been expected. From the fisheries standpoint we have seen the publication of Beverton and Holt's (1957) classic study "On the Dynamics of Exploited Fish Populations", and Nikolskii's (1969) "Theory of Fish Population Dynamics", together with the book on "The Biological Basis of Freshwater Fish Production" edited by Gerking (1967) and Cushing's (1968) "Fisheries Biology". These are all excellent and indispensable works, but they, together with a few others, can be matched by scores of corresponding works on the other animals of major economic significance—mammals, birds and insects. The present work is not an economically based study, but merely an ecological evaluation of growth in fish. It is an attempt to pursue in some depth the essential ecological implications of growth manifested by individual fish in relation to the development and dynamic maintenance of fish in populations. A partial justification for this approach is that, in my experience, "general" texts on animal ecology are usually deficient in their treatment of fish. Many of these general textbooks have been written by ecologists who are primarily ornithologists, mammalogists or entomologists, and their treatments of ecology—even, in many instances, the theories they hold and the definitions they propose—reflect their particular research bias. Sometimes, such bias may be harmless, but the frequently naive viewpoints expressed about the biology of fish can be painful to one who actually works with these animals. This is not a textbook, but an attempt to bring fish to the wider attention of ecologists and students of ecology. I hope it will also be useful to biologists specializing in fish ecology or fisheries research.

I have treated the literature selectively. The citations are confined to studies bearing significantly on the topic or problem and I have not attempted to provide an exhaustive reference list. On the other hand, in a number of instances, the major findings of a particular study are recounted in some detail. The justification of such an approach is that the reader can appreciate more fully the essential lessons of the work and it is easier to comment critically on them.

The book is organized on somewhat different lines from most ecological works in its relatively strong emphasis on "physiological" evaluations. The physiology is very straightforward and includes a little respiration and bioenergetics and a small amount of cell biology and endocrinology. These topics are not infrequently treated in other ecological texts, but have not perhaps been utilized in them in quite the same way. The justification of this approach is explained more fully in Chapter 1 and is implicit in the discussions of physiology and ecology of growth developed later. It stems from my own belief that ecology should *not* be regarded as an area of biology that is becoming increasingly specialized and circumscribed, developing its own special rules and laws—a view that is, so it seems, becoming disturbingly common. My contention is that ecology should provide a programme for solving problems that involve organisms in their biologically proper contexts, rather than being a set of rules to tell how all complexes of organisms function.

This book is an attempt to consider the characteristics of growth in fish, the ecological consequences of this growth and finally to develop a programme for an "ideal" attack on the various facets of fish growth in nature. In a way it may be regarded as an exercise in developing a methodology suitable for the analysis of a complex of problems within a branch of ecology—problems, in which one of the connecting links happens to be growth. Complex models required for the solution of specific ecological problems involving growth are not provided. What I have attempted to do is to present, in some depth and detail, general information about growth biology in fish and other animals and to indicate how models may be introduced where their use is indicated. Readers will notice a bias towards fish populations of inland waters over marine ones. This may seem somewhat odd in view of the preponderance of work on marine fish populations in the world's fishery literature. However, my own work with fish has been in inland waters and I believe that, because of the greater ease with which they can be sampled and the more certain and detailed knowledge of the physico-chemical and biotic environment in inland waters, a good case can be made out to justify a selective emphasis on them.

In the preparation of this book a number of colleagues and friends have helped me in discussion of ideas, in criticizing constructively, in supplying me with valuable information or drawing my attention to relevant items in the literature and in reading various sections of the manuscript at different stages of its development. These people include R. G. Bell, T. Burdon, P. Greenham, B. Harasymiw, J. S. Lake, A. G. Nicholls and Professor J. D. Smyth (in whose Department the book was written), to all of whom I am extremely grateful.

I wish to thank the Leverhulme Trust for the opportunity afforded by the award of a Leverhulme Visiting Fellowship to the University of Singapore from December 1970 to February 1971 to visit fish ponds and fish cultural enterprises in Singapore, Malaysia and Java. My ideas about certain matters raised in Chapters 5 and 9 have been more fully developed because of this opportunity.

I am also grateful to the publishers for their helpful suggestions.

Canberra A. H. WEATHERLEY
January, 1972

Acknowledgements

Grateful acknowledgement is made below of Figures and Tables from other sources which are used in this book: Figs 8.2, 8.3 and Tables 8.1, 8.2, 8.3 by permission of the Director of the Bingham Oceanographical Laboratory, New Haven; Figs 2.6, 3.2, 3.3, 4.1, 4.2, 4.6-4.9, 5.2, 5.3, 7.6, 8.1 and Tables 3.3-3.5, by permission of Blackwell Scientific Publications, Oxford; Figs 2.2, 4.4, 4.5 by permission of Cambridge University Press; Fig. 2.1 by permission of The Clarendon Press, Oxford; Figs 2.7, 3.1, 3.4-3.6, 9.3 and Table 3.1 by permission of the Commonwealth Scientific and Industrial Research Organization, Melbourne; Fig 5.1 by permission of Professor S. M. Das and Dr S. K. Moitra and the National Academy of Sciences of India; Figs 4.10-4.14, 4.21, 4.22, 8.9 by permission of the Fisheries Research Board of Canada from its Journal; Figs 8.4-8.8 and Tables 8.4-8.6 (from "The Horokiwi Stream") by permission of the Fisheries Research Division, Marine Department, Wellington, and Mr K. R. Allen; Figs 6.5-6.9 by permission of the Controller of Her Majesty's Stationery Office; Fig. 5.5 by permission of the Editor of "Nature"; Table 4.1 by permission of Dr Grace E. Pickford and Dr James W. Atz, Bingham Laboratory, Yale University; Fig. 2.8 by permission of Pitman and Sons Ltd; Figs 4.16-4.19 by permission of The Royal Society of Canada and Dr J. R. Brett; Figs 9.1, 9.2 by permission of the Royal Society of Edinburgh and Professor J. E. G. Raymont; Fig. 2.3 by permission of *The Quarterly Review of Biology*; Fig. 2.4 by permission of Charles C. Thomas (Publisher) and Dr N. Shock; Fig. 4.20 by permission of the Editor of *The Transactions of the American Fisheries Society*; Fig. 2.9 and Tables 2.1-2.4 by permission of John Wiley and Sons, Inc.; Fig. 7.5 by permission of the Wisconsin Academy of Sciences, Arts and Letters, Madison.

CONTENTS

7. Predator-Prey Relationships among Fish

8. Production

9. The Trophic Environment and Fish Growth

10. An Operational Programme for the Study of Fish Growth

1 Fish Growth as an Ecological Problem

I. THE DEVELOPMENT OF ECOLOGY

This chapter outlines that view of animal ecology which leads to the particular treatment of fish growth adopted in this book.

Ecology has undergone profound changes in range and scope since the publication of certain pioneering studies which attempted to provide formal syntheses of the ecological facts and ideas of their day (Shelford, 1913; Elton, 1927). Representative surveys of recent ecological thought may be found in Allee *et al.* (1949), Odum, E. P. (1959) and Macfadyen (1963). Some notable attempts to develop rigorous and comprehensive theoretical analyses of certain aspects of animal ecology have also been published (Andrewartha and Birch, 1954; Lack, 1954, 1966). At the moment, however, animal ecology must still be regarded as an "impure" science.

Sometimes, the view is advanced that ecology should, at this stage of its development, move rapidly towards a formal rigour of concept and formulation (Sears, 1964; Odum, E. P., 1964; Blair, 1964; but see Engelmann, 1966). But I think the rigour and coherence that ecology should aim for is akin to that of engineering or systems analysis rather than of physico-mathematical sciences (see Hughes and Walker, 1965; Watt, 1966, 1968; Weatherley *et al.,* quoted in Myers, 1967).

Allee *et al.* (1949) wrote of ecology in a way of which most contemporary ecologists would still approve: "Ecology may be defined broadly as the science of the interrelation between living organisms and their environment, including both the physical and biotic environments, and emphasizing interspecies as well as intraspecies relations."

Such a definition can still be used as a working approximation, though like other attempts to epitomize areas of science whose problems are very complex and diverse, it cannot offer more than a hint of the full scope of the subject.

II. SOME APPROACHES AND PERSPECTIVES IN ANIMAL ECOLOGY

We now consider briefly some of the main viewpoints that influence the ways ecologists go about their work.

A. The Population

One of the origins of animal ecology was Shelford's (1913) early account which attempted to explain distribution mainly as the result of different tolerances and responses of organisms to the physico-chemical conditions of the environment. Much modern work by so-called *autecologists* is derived from this conception and it has produced many valuable discoveries. It has, however, also tended to be a somewhat static approach, rather underemphasizing the need for investigating an animal's entire life cycle—the different stages of which often have quite different tolerances and environmental conditions. Moreover, the approach originated before ecologists had fully grasped the importance of animal behaviour in determining distribution.

Population dynamics, a discipline owing much of its origin to C. Elton (in common with many branches of ecology), began to develop rapidly about 40 years ago, to become eventually a powerful analytical method in the hands of certain investigators, characterized increasingly during recent years by a complex mathematical content (e.g. Andrewartha and Birch, 1954; Ricker, 1954, 1958; Beverton and Holt, 1957). However, a number of important contributors to population ecology have fortunately contrived to present many of their investigations and ideas in forms intelligible to non-mathematical biologists (Allee *et al.*, 1949; Andrewartha and Birch, 1954; Cole, 1954; Lack, 1954, 1966, 1968; Slobodkin, 1962; Nikolskii, 1969).

A major concern of population dynamics has always been to determine, by experiment and observation, those parameters of populations dealing with natality and mortality, so as to be able to assess the capacity of populations to increase numerically with time. In a sense, all modern quantitative population studies derive from Malthus' essay on increase in the human population. His intellectually lineal descendants—human demographers, biostatisticians, actuaries and census takers—have all influenced the development of population dynamics, especially its theoretical aspects.

Many animal populations exhibit great numerical changes in the course of time and their ability to survive these changes is one of the more remarkable facts that this type of ecological investigation has revealed. The analytical study of these fluctuations in numbers is fundamental to the understanding of how populations regulate themselves, or are controlled or regulated by their environment, and is a primary tool of the applied ecologist who wishes to control animal populations or rationalize their economic exploitation.

B. The Community

The community concept has been recurrently important in ecological thought for a long time, as Allee *et al.* (1949) have shown, and it offers both a good example of a theme demonstrating current problems of ecological analysis and a convenient departure point for an explanation of the approach to ecological problems adopted in this book.

The following is a serviceable definition of the community: "... the major community may be defined as a natural assemblage of organisms which, together with its habitat, has reached a survival level such that it is relatively independent of adjacent assemblages of equal rank; to this extent, given radiant energy, it is self-sustaining" (Allee *et al.,* 1949).

Additional arguments that follow this definition are that communities are composed of "ecologically compatible species populations", that they tend towards ecological stability as they evolve towards the "climax" state, that they have characteristic appearances (at least to the trained observer) and that because they create their own internal environmental conditions they tend to select the types of organisms that can enter and survive in them.

Elton (1966), following some 40 years' study of communities, has emphasized the difficulty of establishing a community's boundaries by pointing out that "... no habitat component with its animal community is a closed system. Though for many practical and working purposes the structural boundaries are highly useful and informative they are constantly being passed by population movement . . ."

Elton added that the community is indeed an open system and that in some ways it could be useful to compare it with the "field system" of the physicist ". . . with lines of influence spreading from an active centre . . ." He also discussed the interlocking nature of community units, indicating that though the most intimate relationships were usually with adjacent units this was not always the case.

Williams (1967) has described how mathematical and computer analysis may help to delineate breaks and discontinuities between communities, although emphasizing that this can only improve the definition of limits which must first be discerned and characterized intuitively.

C. Community Energetics

Some ecologists have argued forcefully that there are enough parallels between the functional organizations of single organisms and communities to regard the latter holistically as "supra-organisms" (Allee *et al.,* 1949). Such views have stimulated recent investigations of production, metabolism and energy flux of communities. For many years physiologists have been measuring the total metabolic exchange of animals—especially mammals—using such methods as direct calorimetry, indirect calorimetry based on measurement of respiratory exchange and direct measurement of food rations consumed. Conversion factors of appropriate value make feasible interconversions between data on food, respiration and heat flux (Brody, 1945; Kleiber, 1961; and Chapters 2 and 4).

Bertalanffy (1951) has stated the main justification, as regards bioenergetics, for treating communities and other complex associations of organisms in a holistic manner: as it cannot be hoped to measure energy fluxes of each individual organism in a community, it might be an acceptable alternative to measure their combined metabolism. It is claimed that such a procedure allows us to make quantitative comparisons between assemblages of organisms for particular purposes.

Certain ecologists have actually treated communities as supraorganisms and tried to measure the energy flux or production process of the whole; at other times they have tried to distinguish various lines of energy flux between different trophic levels or through feeding webs. The essential principles and some examples are given in Macfadyen (1963) and Odum, E. P. (1959).

III. COMMENTARY ON CURRENT ECOLOGICAL THOUGHT

The dynamic, complex nature of communities will probably continue to make it difficult for ecologists fully to render adequate accounts of them and the difficulties are greater for animal, than for plant communities. Though some animals are sessile and others so sedentary (or so small) that their entire life cycle may be completed within a few square centimetres or millilitres, many others are either capable of individual movement, are carried passively or are conveyed within the bodies of parents or hosts. Some animals may inhabit a region for years, then desert it suddenly for better food or shelter elsewhere–and not necessarily merely because the former habitat has become intolerable.

It is because "the structural boundaries" of communities are "constantly passed" that animal ecologists may have to reserve final judgement on the validity and universality of the community concept. And it does not really help to solve this dilemma by comparing certain community attributes with those of physical fields. The latter arise as the result of forces or influences that vary in intensity in an essentially simple manner in space and time. Communities are not simple either in the arrangements of their material components or the channellings of matter and energy within them.

Ecologists cannot, however, object to the employment of the term community as a very useful operational concept, although to apply it in a more rigid sense would impede certain ecological analyses proposed in this book.

Specific mention of the meaning and derivation of the term ecology usually occurs early in books on the subject, but it may be an advantage to avoid defining the subject more closely than was done by Allee *et al.* (1949) over 20 years ago. It seems more important to emphasize that ecology today encompasses the areas of study, problems and methods of a certain group of biologists, so that attention may then be directed to the rapidly changing present nature of their work.

Among the most important current topics are analyses of the ecological differences between the different stages of life cycles (Cole, 1954; Bonner, 1965; Weatherley, 1965) and of mechanisms whereby populations maintain themselves within certain numerical limits (Allee *et al.*, 1949; Andrewartha and Birch, 1954; Ricker, 1954; Beverton and Holt, 1957; Cushing, 1968). Ecological methods already include a heterogeneous assemblage of knowledge and techniques from other branches of science ranging from physics and mathematics to behaviour and sociology, and the list is growing rapidly.

A major aim of modern ecologists is how to obtain better understanding of entire, functioning organisms living in the natural world. It can therefore be confidently claimed that ecology has a central rather than a peripheral role to play in the development of biology. Such a view of ecology means that such varied aspects of biology as metabolic pathways, details of structure or morphology, genetic mechanisms and the analysis of behaviour, together with a miscellany of other matters, may all be considered relevant by the ecologist when they assist him to plot the dynamic status and interrelations of organisms both as individuals and in populations.

Allee *et al.* (1949), following their definition of ecology, added that although organisms, singly and in groups, "are the essential biological units in ecology", we should "exclude the intra organismal or cellular environment except as special circumstances demand its investigation." However, ecologists are nowadays realizing that many circumstances once thought of as special have quite wide application.

Suppose, for instance, that study of the population dynamics of a particular species demands knowledge of its reproductive success in relation to climate. Now, fecundity might be closely correlated with environmental temperature, with secondary effects of humidity and day length also apparent. From such relationships it would often be possible to make multifactorial analyses of fecundity and environmental factors, which would suffice for some computational requirements. But for deeper insight into the breeding biology of the species special physiological, behavioural and possibly developmental studies might be needed. Such detailed investigation is sometimes the only way of obtaining the predictive ability and deep knowledge of the effects of the environment on the organism needed for fully satisfactory ecological work. Such ecological uses of physiology and behaviour are becoming increasingly common for a range of animals, especially mammals (e.g. Myers, 1968; Main, 1968; Brown, G. D., 1968), birds (Marshall, A. J., 1960, 1961), insects (Hughes, 1968; Bartell, 1968; Barton Browne, 1968) and fish (Chapters 2 and 4).

IV. SYSTEMS ANALYSIS IN ECOLOGY

In contrast to the population ecologist or the community ecologist, both of whom have particular specialist viewpoints concerning the structure of associations of organisms, the "systems" ecologist attempts to bring the methods of the systems analyst to bear on ecological problems (Van Dyne, 1966; Watt, 1966, 1968). While it is frequently implied that this represents an almost completely new approach to ecology this is not strictly correct. The comprehensive evaluation of many aspects of certain aquatic ecosystems for instance, in relation to particular research questions, has been attempted before (Chapters 9 and 10). The novelty of the new use of systems analysis in ecology is in the quite conscious attempt to employ the fully developed strategy of this method as it has already been used in the study of weaponry and arms, certain branches of economics and business, engineering, medicine etc. To do this, not only is it necessary to employ superior methods for the collection and recording of data, but also to have standardized procedures for the arrangement and testing of interrelational ideas about the structure of the "system".

The value of systems analysis for ecology and a listing of its requirements can be found in Van Dyne (1966) and Watt (1966, 1968).

Van Dyne emphasized the resemblances between "systems ecology" and "communities ecology". He even admitted that systems ecology is perhaps not really an independent science, "but a point of view."

The necessary features of the scientific methods of the systems ecologist will include:

(i) Skill in formulating problems, without which the elaborate studies that may be required will be ill-conceived or irrelevant.

(ii) Ability to construct sound hypotheses for crucial testing.

(iii) Appropriate use of logic and statistics in evaluating experiments.

(iv) Ability to construct theoretical models (which may be of many types).

To these basics may be added an extensive list of other requirements including skilled use of apparatus for massive data collection and storage, environmental monitoring and access to suitable computers. Work in systems ecology almost automatically assumes a team effort.

Van Dyne (1966) claimed that typical "applied" ecologists were usually likely to be closer than typical "academic" ecologists to the systems ecologist in outlook. He explained that an applied ecologist, such as one responsible for the management of a trout population, was more likely to appreciate the effects on a particular animal population of manifold changes—possibly induced by man—of the watershed, such as grazing, lumbering, mining, road-making etc. The applied ecologist would be more likely to possess an appreciation of at least the general nature of the effects of these factors on the population. Circumstances would also compel him to communicate with hydrologists, botanists, soil scientists, foresters etc.—something not always necessary for (and sometimes actively avoided by) the "academic" ecologist. And all these pieces of information and appreciation of their meaning would form a background to the applied population dynamics of the trout management work, which include judgements on artificial versus natural stocking, bag limits and closed seasons.

Van Dyne was, however, too sweepingly critical of non-applied ecologists. One cannot ignore, for example, the extensive work of C. Elton and his colleagues at Wytham Woods in studying community dynamics. On the other hand, it is true that applied ecologists cannot freely choose their problems from an unlimited range, finding one that nicely fits their temperament and ability. Instead, they must deal with whatever difficulties are peculiar to their research problem.

Both Van Dyne (1966) and Watt (1966, 1968) have emphasized that very high levels of training will be required of those who comprise future teams of systems ecologists. Perhaps, however, a warning note should be sounded here. Williams (1967) strongly defended the usefulness of those botanists able to discern plant communities intuitively and broadly characterize them, leaving detailed analysis to those employing more elaborate methods. It might be added that the most able leaders of teams of systems ecologists might not infrequently have less training than most in special skills or knowledge—i.e. they might be the types of biologists (admired by Williams) capable of identifying problems while maintaining a broad yet vital overview. Considerations such as these should not be ignored when personnel for research teams are being selected, otherwise able workers may be misapplied or even excluded from ecological projects to which they might valuably contribute.

V. THE "PROBLEM-ORIENTED" APPROACH TO THE ECOLOGY OF FISH GROWTH

If the application of systems analysis to ecology becomes more general it may offer a unique way to solve otherwise intractable problems, e.g. the ecological implications of fish growth.

Despite Van Dyne's preoccupation with ecological systems (often whole ecosystems) his approach is problem-oriented in the sense that he insists ecologists should remember that their problems have broad and complex frameworks of which the structure must often be investigated. Therefore the status of a problem has to be understood within its setting.

Growth is a property especially appropriate for systems analysis. In fish, growth is very labile, influenced by food, space, temperature and other factors. Furthermore, since fish are both poikilothermic and live permanently immersed in water, they are very directly affected by changes (e.g. in temperature) in their ambient medium. Since fish affect the trophic status of other organisms in their environment and alter the condition of water in various ways—especially when their population density is high—the relations between organism and environment are far from unidirectional. A study of growth in fish offers a systems ecologist many opportunities to enquire into the dynamic balance and states of change in aquatic ecosystems.

In this book, the attempt is made to provide information and ideas about fish growth as a suitable base for systems studies.

Chapter 2 gives a brief general exposition of various ways of viewing the general biology of growth. Subsequent chapters break the problem of fish growth down in various ways. Chapter 3 discusses the methodology for accurate ageing and growth determination of fish in populations. Chapter 4 is a development of Chapter 2 but deals specifically with fish growth. Chapters 5 to 8 then deal in order with competition and growth in relation to the niche concept, the role of growth in population structure and maintenance, predator-prey relations and production. Chapters 9 and 10 indicate how present knowledge of growth could be used to obtain more detailed understanding of the ecological "status" of growth and strategies are outlined for investigations of several different aspects of growth.

The book suggests first how growth may be analysed, either in the individual animal or in the population. Such analysis has no essential limit because it can be carried to any requisite level of refinement. Secondly, it demonstrates that growth in size or numbers is an inevitable accompaniment of the life of the fish or the population of fish. Even the condition of "steady state" in a population (Chapter 6) is a dynamic condition and the result of a nice balance of gains and losses of individuals. It can be turned to overt growth merely through temporary suppression of mortality. Similarly, non-growth in the individual can give way to growth if the balance between anabolism and catabolism be changed. Finally, the book attempts to trace the relation of growth and population dynamics. It is assumed that each aspect of this relation can in principle—and will eventually in fact—be quantitatively accounted for.

If the book succeeds in drawing the attention of more people to the possibilities for research in this area and furnishes some material for teachers of animal ecology, whether or not they are concerned directly with fish population dynamics, it will have fulfilled its principal purpose.

2 | Growth in Animals: Central Concepts

This chapter briefly introduces the general theme of growth in animals. It is included because many ecologists working on growth problems tend to confine their ideas on the subject to an arbitrary and undesirably narrow minimum. It therefore seems desirable that they should realize from the outset of their studies that though certain growth problems may have peculiar and specific features, others will be inevitable results of general rules of growth.

D'Arcy Thompson (1942), an important pioneer of growth analysis, declined to define growth rigidly, evidently believing that to do so might artificially restrict the comprehensive approach he was seeking. Other leading workers have been similarly heuristic; e.g. Bertalanffy (1960) wrote, "A definition of growth can only be of an 'operational' nature, and its value is exactly equal to what it may offer for a specific problem of research."

In this book *change* in size (length, weight, bulk) with time, and change in numbers with time in the case of a population, will be what is usually intended by use of the term "growth". Generally, we shall be dealing with quantitative trends in size or numbers of organisms when size or numbers are tending towards some upper limit set by the type of organism or by the external or internal conditions of its life. We should note that both individuals and populations "may grow larger or smaller" (D'Arcy Thompson, 1942) in terms of either their combined mass or their numbers.

I. GROWTH CURVES

D'Arcy Thompson (1942) and Medawar (1945) outlined various ways in which growth of organisms (including fish) could be graphically depicted, and the growth curves in Fig. 2.1 illustrate "the six chief integral and derived curves of growth; from them the majority of our useful inferences about the nature of growth can be drawn." In ecology the curves of growth, growth rate and specific growth rate (Fig. 2.1a, b, e) are probably the most useful generally.

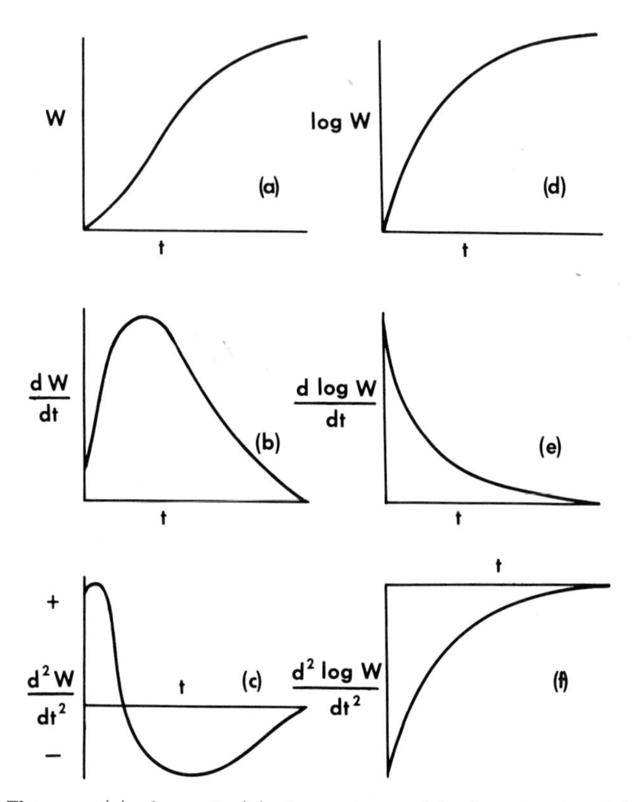

Fig. 2.1. The curve (a) of growth; (b) of growth rate; (c) of acceleration; (d) of specific growth; (e) of specific growth rate; (f) of specific acceleration. The curves have been plotted from an equation for the Gompertz function, but the scales of the ordinates are adjusted to make the height of each graph uniform. (After Medawar, 1945.)

Curves of the type in Fig. 2.1a demonstrate very generally the course of increase in size (length, stature, weight, volume) of many individual organisms, or of numerical increase in populations. D'Arcy Thompson (1942) gave numerous examples of this curve of growth, e.g. the human population, the beanstalk, lupines, yeast populations.

Figure 2.2, based on Thompson's graph of growth in the beanstalk from Sachs's data, could as easily have been the curve of growth for other plants, various animals or for cultured microbial populations. This so-called logistic (S-shaped) curve also appears in

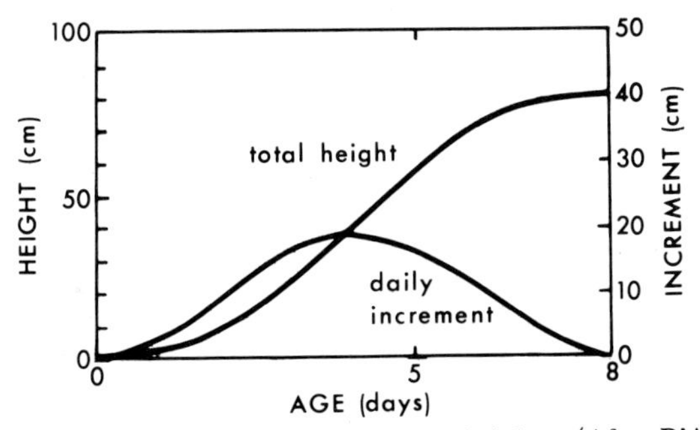

Fig. 2.2. Growth in height of a beanstalk, from Sachs' data. (After D'Arcy Thompson, 1942.)

Fig. 2.3 with the derived curve based on the increments which account for the shape of the S-shaped curve. The extent by which the shape of a logistic departs from complete symmetry will be reflected in the degree of skewness shown by this derived "Gaussian" curve (D'Arcy Thompson, 1942). This curve is also that of "growth rate" as given by Medawar (1945), Fig. 2.1b. The bell-shaped "Gaussian" curve closely resembles the distribution

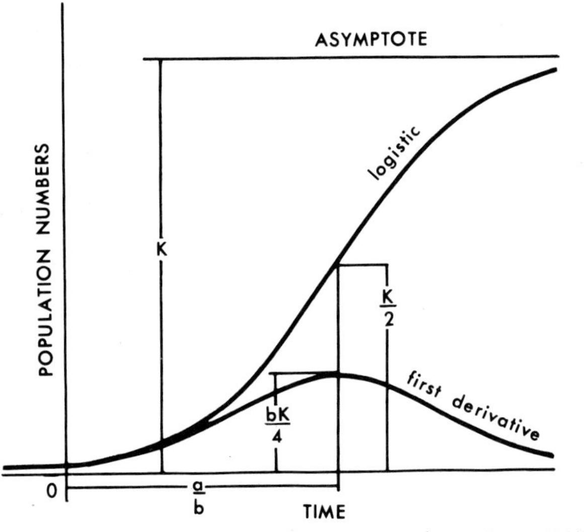

Fig. 2.3. The logistic curve and its first derivative. (After Pearl, 1927.)

curve used to express frequency of occurrence of a range of values of some attribute (say, length or weight) in a "normal" population. Even in that case, the values the "Gaussian" curve expresses may be "transformed" readily into an S-shaped curve of distribution, in which the proportional frequency of occurrence of, for instance, a given length in the population is plotted against that length.

The rapidly rising section of the S-shaped curve signifies exponential growth, either that produced by increase in the number of individuals in a population or in size or bulk of an individual organism. It is a decline in specific growth rate (Fig. 2.1e) which accounts for the S-shape of the summative or logistic curve of growth including the presence of the upper asymptote. Experiments with populations of micro-organisms and cultures of tissue cells suggest that, under conditions in which factors such as space and nutrient supply are not limiting, growth might be held indefinitely in an exponential phase.

The formula for exponential growth is

$$\frac{\Delta y}{\Delta t} = \mathrm{r}y \tag{2.1}$$

where $\Delta y/\Delta t$ is rate of change in size or numbers per unit time, r is the exponent for rate of increase in size or numbers ($\mathrm{r} = \Delta y/y\Delta t$) and y is the size of the growing organism or the number of individuals which is increasing. In the latter case, the integrated form of (2.1) applies:

$$Nt = No\mathrm{e}^{rt}$$

where No is the number at the beginning of the observation period, Nt the number after time t and e is the base of natural logarithms. What r actually represents is the difference between specific birth and death rates in a population.

It is the general form of such an expression that is of present interest. An organ (composed of cells) or a whole organism (composed of collections of cells) could display growth of a generally multiplicative or exponential character as readily as a population of individuals.

Early experiments that seemed to indicate potential immortality for fibroblasts *in vitro*, if culture conditions were satisfactory, may have been misleading. Chick fibroblasts will not survive long in culture. Mouse cells may divide indefinitely, but apparently only because they undergo spontaneous transformation *in vitro*. HeLa cells, derived from human cells, have formed an unbroken culture line since 1952, but they were abnormal from the beginning, having from 50 to 350 chromosomes per cell instead of the normal human complement of 46. Indeed, it appears likely that only *transformed* higher cells (e.g. cancer cells) will continue to divide indefinitely *in vitro* when continually subcultured into new media and containers. Hayflick (1966, 1968) has indicated (Fig. 2.4) the usual situation that follows when *normal* cell lines from higher vertebrates are started.

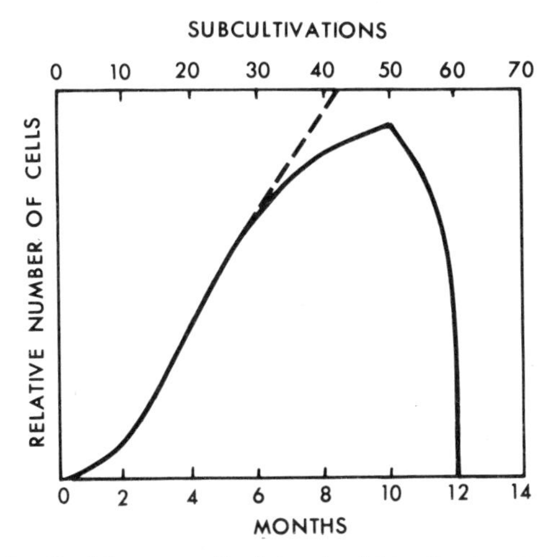

Fig. 2.4. "Lifetime" of human cells determined by allowing population of cells to multiply until it had doubled. After a culture of cells from embryonic tissue had grown to a particular size it was divided into two. All division ceased after about 50 such subcultivations had doubled. It is possible (though rare) for a spontaneous change to occur at any time, after which the cells may multiply indefinitely (broken line). (After Hayflick, 1968.)

In vitro demonstrations, such as regrowth of repeatedly ablated liver tissue (Goss, 1964), suggest that many organs would continue to grow in an exponential manner, only for limits imposed by their biotic environment. Needham (1964) pointed out that variability of the shape of the sigmoidal growth curve indicates that there is no close, or absolutely determined, relationship between the initial high growth rate in populations or individuals and its subsequent decline; and this is "what might be anticipated if, for instance, the one is an intrinsic potentiality and the other an imposed control."

The derivation of the logistic from the exponential form of curve follows if growth gradually declines from an earlier multiplicative phase and tends towards an asymptote.

Thus $$\frac{\Delta y}{\Delta t} = ry$$

(where y has a definite upper limit reached abruptly when growth ceases), but in the logistic

$$\frac{\Delta y}{\Delta t} = ry \frac{(k - y)}{k} \qquad (2.2)$$

where k is the upper asymptote.

The integrated form of this equation is

$$y = \frac{k}{l + e^{a - rt}} \qquad (2.3)$$

where e is the base of natural logarithms and $a = r/k$. This is the form which Pearl (1927) gave in describing the growth of a yeast population, and in that case

$$y = \frac{66{\cdot}5}{l + e^{4{\cdot}1896 - 0{\cdot}5355x}}$$

the values of the parameters being determined empirically.

Slobodkin (1962), Odum, E. P. (1959) and Andrewartha and Birch (1954) all warn against an uncritical use of logistic (or other) curves, in so far as growth of population numbers is concerned; Odum, E. P. (1959) stressed (pp. 185-186) that there were many equations which would produce sigmoid curves and that "mere curve-fitting is to be avoided. One needs to have evidence that the factors in the equation are actually operating to control the population before an attempt is made to compare actual data with a theoretical curve." Andrewartha and Birch (1954), in their detailed critique of the use of the logistic curve, noted that it was inappropriate to apply the curve to growth of populations which did not initially possess a stable age-distribution,† which is a basic premise of success in fitting the curve. They concluded (p. 385 et seq.) that "despite its theoretical limitations, the logistic curve remains a useful tool for the ecologist. Because of its limitations, too much reliance should not be placed on it in particular cases, until it has been verified empirically for each case."

Not infrequently, neither the logistic curve nor any of its approximations is adequate to describe growth of either

† According to Andrewartha and Birch (1954), "This is the age-distribution which would be approached by a population of stable age-schedule of birth-rate and death rate . . . when growing in unlimited space."

populations or individuals. Andrewartha and Birch (1954) give examples of what they regard as serious departures from the logistic form of growth for animal populations and Watt (1955) has given examples of the now well known "crashes" that can occur in populations.

As for growth in individual organisms, *Spirogyra* has an extremely uneven rate of growth in length. Animals such as tadpoles and silkworms grow slowly at first, then exponentially, but their size subsequently declines precipitately, matching an abrupt change in form, composition or bulk of the body (D'Arcy Thompson, 1942). In man and shark, sigmoidal curves adequately describe foetal growth but not growth after birth (D'Arcy Thompson, 1942; Olsen, 1954). Many arthropods grow fairly steadily during development to reach a definite final size quite abruptly, without any appearance of an asymptote. Other arthropods, with more rigid exoskeleton, have a growth pattern that bears a stepwise correspondence to the number of moults. In these insects weight increase is sometimes out of phase with length increase, because the latter can only take place between moults, while weight increase is not so constrained (D'Arcy Thompson, 1942; Bertalanffy, 1960; Needham, 1964).

Of course, the fact that so many animals have complex life cycles makes it very unlikely that the growth of all life stages could be combined into a simple curve—such as the logistic. Often, too, the pattern of growth characteristic of a particular species is seen best in sequential observations of individuals rather than in mass statistics. In human growth, for instance, the variable onset and duration of the adolescent growth spurt blurs the essential configuration of growth during this key period if data from many individuals are combined into a general curve (Tanner, 1955, 1960).

On the other hand, the overall growth patterns of many species in nature appear almost immutable. Mammals and birds of particular species are notably constant as regards their final body size. Yet, many recent experiments on pigs, rats and poultry have demonstrated that under carefully controlled experimental conditions these animals can be made to remain stunted for a long period by no other means than severe dietary restriction. When fed fully these artificially deprived animals can grow to full-sized, normal adulthood; although in rats, food deprivation must be preceded by normal nourishment at least to nine weeks old, if subsequent recovery is to be complete (McCance and Widdowson, 1962; Widdowson, 1964; Lister, Cowan and McCance, 1966).

That rehabilitated pigs tend to be more obese than normally fed littermates led Lister and McCance (1967) to comment that age undoubtedly influences the metabolic fate of consumed food, which probably explains why rehabilitated animals are fatter, because when growing over the same weight range they are a year older than normal controls. They also pointed out that completeness of rehabilitation after deprivation in rats, guinea pigs, pigs and poultry, depends on the deprivation having taken place before the age at which animals lose their capacity for growth.

Fish growth is, like that of many plants, essentially indeterminate. It is therefore difficult or impossible to judge a fish's age from its size, unless one has other criteria (see Chapter 3). Figure 2.5 shows growth in length of trout which, at four years, may vary from 20 to 60 cm in different populations. This represents approximately a 27-fold weight difference! Figure 2.6 further shows the nature of the flexibility of fish growth, in that tench were taken from a wild population of the species and put in farm dams in which—due to more optimal feeding conditions—they were able to grow as much, in less than one year, as they could in four years in their wild environment. It should be noted that though intraspecific differences in growth rate between populations

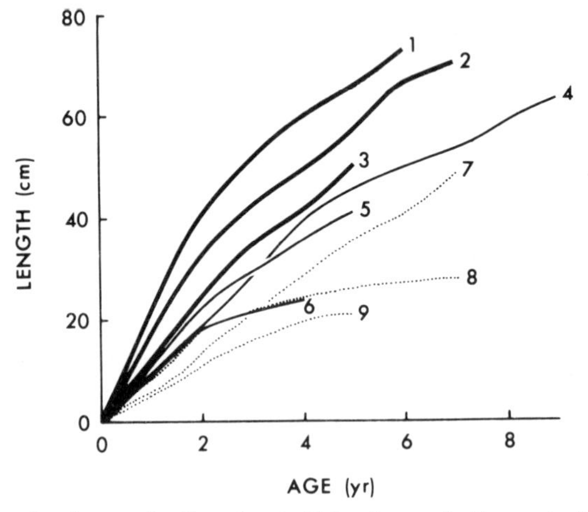

Fig. 2.5. Growth of trout in New South Wales (▬▬), Tasmania (———), and the English Lake District (· · · · · ·). The numbered curves are based on the work of the following authors: 1-3, Lake (1957); 4, Nicholls (1961) for Plenty River fish; 5 and 6, Nicholls (1958b) for Meander and St. Patrick's River fish respectively; 7, Allen (1938) for Windermere trout; 8, Frost and Smyly (1952) for trout from Three Dubs Tarn; 9, Swynnerton and Worthington (1939) for Haweswater trout. (After Weatherley and Lake, 1967.)

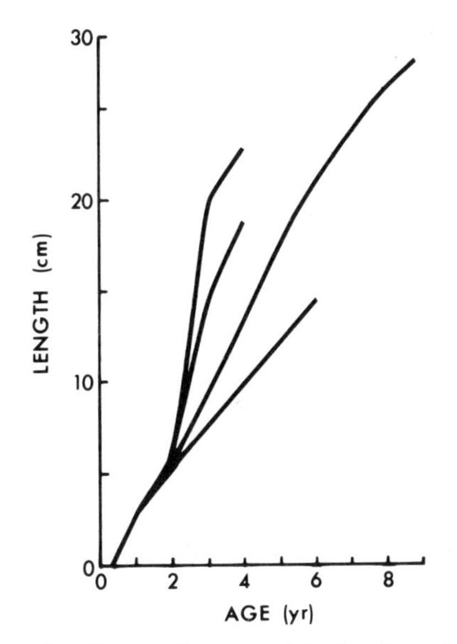

Fig. 2.6. Mean growth, all year-classes combined, of tench from four Tasmanian environments. The lowest curve represents growth in Lake Tiberias, the next from the Coal River. The two highest curves represent growth in farm dams. The tench were placed in the dams from Lake Tiberias when about 2 years old. Curves display the labile growth rate of tench. (After Weatherley, 1959.)

can be very large, differences between members of a single population are usually not nearly as great.

Relatively little is known concerning growth among the other lower vertebrates—the amphibians and reptiles—a deficiency attributable to their lack of economic importance. However, many species of reptiles do exhibit a considerable intraspecific range of growth rates, over which climate exerts a direct or indirect influence. The extreme flexibility characterizing fish growth seems less common among reptiles, though critical evidence is wanting. Anuran amphibians have a two phase life cycle and tadpole growths may vary considerably, apparently depending mainly on a food or space factor or a temperature factor (Savage, 1961). But, although age at metamorphosis may vary, *size* at metamorphosis is usually sufficiently species-specific to be a useful taxonomic feature. Anurans with infra-oval development of the tadpole have less variable growth rates, because food and space available to the larva are fixed. Post-larval growth has been little studied.

In general, the growth in nature of other lower vertebrates appears to show less variability than that of fish but rather more than that of birds and mammals.

Sigmoid curves may be suitable for portraying the growth of a variety of organisms, but their applicability to fish growth patterns depends on the particular case. They may be adequate in depicting the growth of fish as the outcome of "an intrinsic potentiality and . . . an imposed control" (Needham, 1964). Growth of salmon in rivers (Allen, 1940, 1941) and individual trout in small hatchery troughs (Nicholls, 1957) could be satisfactorily portrayed by such curves. Indeed trout showed successive effects of springs and summers during which temperatures were not limiting and growth was rapid, followed by winters during which growth virtually ceased for several months as Fig. 2.7 indicates.

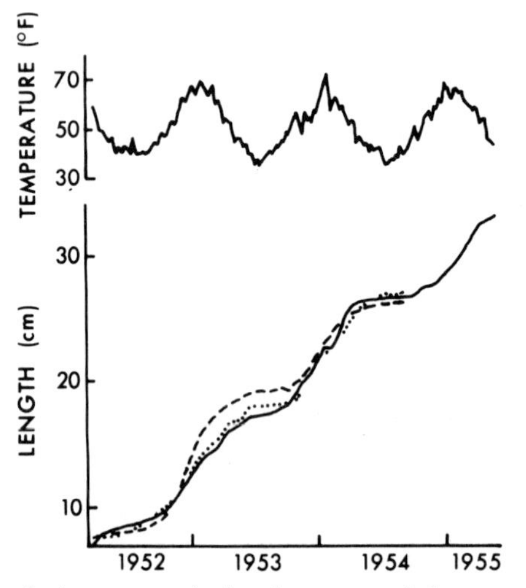

Fig. 2.7. Growth of three trout in hatchery, recorded at approximately 4-weekly intervals. —— Brown trout; .... male rainbow trout; – – – female rainbow trout. Mean weekly temperatures are also given. (After Nicholls, 1957.)

The size-relation analysis of the body parts of a growing organism is one aspect of the mathematical approach to the description of growth. Huxley (1932) developed and elaborated this method, which was originated by D'Arcy Thompson. Huxley sought to show that many "problems of relative growth" were resolvable into essentially simple heterogonic relations. As is mentioned in Chapter 4 this type of relation is expressed as

$$y = bx^k,$$

the formula referred to by Huxley as the simple "heterogeny formula" in which "the constant b is here of no particular biological significance, since it merely denotes the value of y when $x = 1$, i.e. the fraction of x which y occupies when x equals unity." Huxley's analyses have allowed at least partial interpretations of a variety of puzzling growth phenomena. He showed that, starting with a certain ground plan, constant differences in growth rate for different organs, or for different dimensions of the same organ, are often enough to produce bizarre departures from that plan. Sometimes sudden but quite simple shifts in the value of the allometry constants can account for even more baffling changes in growth pattern. Huxley was able, at the particular level of analysis on which he concentrated, to elucidate growth patterns of a wide variety of organisms and their parts, suggesting that solutions of many problems of growth depend on the identification of "morphogenetic fields" of influence and the like. This latter term of Huxley's still indicates a general approach favoured by some functional morphologists.

II. METABOLISM AND GROWTH

Huxley (1932) stated that:

> the law of constant differential growth-ratio . . . teaches us the striking fact that relative growth-rates of different parts of the body may stay constant over long periods of growth, which clearly is important as a contribution to the problem of form co-ordination and the orderliness of form-change, but it sheds little light upon any aspect of the growth-process itself.

Apart from the mathematical formulation of growth as an additive phenomenon another valuable quantitative approach is to account for the matter and energy transactions by which growth may be represented. We must bear in mind that part of the food consumed by animals supplies the energy needed for the search for food itself, part goes to "standard" or "basal" metabolism, another fraction to tissue repair, some is excreted and some, altogether rejected by the body, is voided as faeces. Only that remaining is available for actual growth. Accounting such as this can be complicated and difficult to apply to wild populations of mobile animals (Weatherley, 1966 and see Chapter 4), but worth the attempt because of the quantitative insight that may emerge about the life processes of the animal in relation to its growth.

If growth is as central to the ecology of a species as has been suggested, then it should be evaluated most appropriately in terms of its total "cost" to the ecosystem. How much matter and energy are channelled into it? At what rate are they supplied and how does this vary at different times of the year? We are led also to enquire how many organisms are using matter and energy at the calculated rate. At what rate is material from dead animals returned to the ecosystem for re-use? Or, if animals are preyed upon and therefore permanently removed from the ecosystem, at what rate does this occur and how much loss of matter and energy are involved? Such questions are central to much work now being done by general ecologists. Our present circumscribed interest in them concerns merely growth.

There are many ways to investigate the bioenergetics of animal growth. In animal husbandry the advantages of knowing the manner and magnitude of the partitioning of matter (and energy) into the several competing demands of growth, maintenance, activity, excretion etc. have been fully described by Kleiber (1961) and Brody (1945).

To illustrate the application of this approach, two organisms of different species living in the same ecosystem are considered. Their form and habits may be quite different but if they have similar demands for food they may be affecting the ecosystem in very similar ways. This similarity of effect will increase in proportion to the extent to which their food habits converge, as will be more fully discussed later (see especially Chapter 5). The present point is that if two apparently different animals are removing material of the same kind from the one ecosystem, a change in the food supply may affect both animals, though the detailed consequences of such an effect may depend on the particular requirements for food materials in different animals. And in cases of partial or temporary starvation, the course of "degrowth" in different animals could be expected to be different (Table 2.1).

Most ingested materials are complex mixtures of variable proportions of a wide range of carbohydrates, fats and proteins. Included with these are varying quantities of minerals, vitamins (both usually small in actual bulk), plus indeterminate amounts of undigestible fibre, chitin, fur, feathers, scales, bones etc. It is therefore frequently convenient to think of food and its metabolic fate in terms of energy equivalents. Moreover, though the animal encounters a great deal of variety in its food, it is an integrating system which is able to convert this variety into final body substances that remain typical in composition during much of the animal's life.

Table 2.1
(After Kleiber, 1961)

Relative weight loss of animal organs during starvation
(Expressed in percentage of original weight)

Organ	% Weight Loss Pigeon (Chossat)	Cat (Voit)
Fat tissue	93	97
Spleen	71	69
Pancreas	64	17
Liver	52	54
Heart	45	3
Skeletal muscle	42	31
Intestinal tract	42	18
Skin and hair	33	21
Lungs	22	18
Brain	2	3

Needham (1964) has provided a simple scheme (Fig. 2.8) demonstrating that much less than half of the food energy ingested ends up as new tissue or the products of tissue. A considerable fraction is usually undigestible, while the effect of the specific calorigenic action of foods (especially proteins), plus the loss in nitrogenous excretion, removes or dissipates perhaps a third of what is digested. Needham's (1964) scheme was much the same as

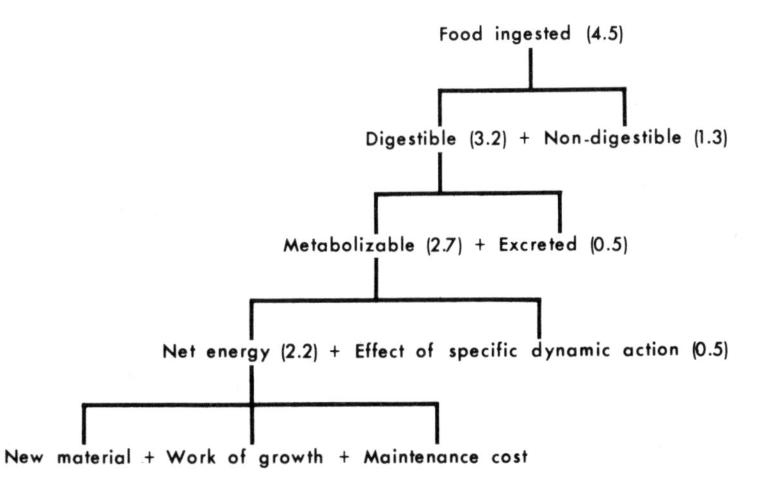

Fig. 2.8. Scheme of food utilization in animal. Values assigned to variables merely indicate typical orders of magnitude. (After Needham, 1964.)

Kleiber's (1961), except that Kleiber's terminated at *net energy*, defined as "the amount of energy which appears as the saving of body substance or as the animal product (body fat, body protein, milk, eggs, wool or work)." Since *net energy* is the only energy that the body can actually tap for all its vital needs those needs will inevitably compete with each other. If the need for new material is high the work of growth will almost certainly also be high. Moreover, *maintenance cost* will always cover the inevitable, minimal catabolism required to keep the vital processes functioning, even in the absence of all overt activity. But in ordinary life the cost of a workload will almost always be superimposed on this. This workload will be incurred in the normal activity of everyday life in wild animals—which will include a certain amount of undirected or "random" movement; the energy required in the search for food itself; the catabolic cost of avoiding predators or potential competitors, of constructing and maintaining shelter, of indulging (in most cases only seasonally or periodically) in courtship and mating. Only the most generally quantified scheme of the way in which net energy is shared between the conflicting demands of the organism should therefore be proposed. Each population will represent a separate case requiring individual computation.

The energy balance sheet implied by the Needham-Kleiber scheme (Fig. 2.8) may have one of the two terms in each of its successive stages computed by difference. For instance if it is known in stage three of the scheme how much of the food energy is excreted, by subtracting this from what has been digested (stage two) we have the energy that remains for further metabolism. Calculations can, however, be more complicated. The specific calorigenic action of foods can, in principle, be determined for any animal; in practice, especially for small aquatic animals such as many fish, there are formidable technical problems to overcome. However, assuming that net energy is obtainable as a first approximation, there remains the problem of measuring its partitioning. Where surface materials (wool, hair, feathers, scales, skin) are not being sloughed off at any significant rate and where energy is not being released as reproductive products, milk etc., in other words where all body product is growth, such product may be determined simply by weighing the animal at known intervals. Simple calorimetry enables us to convert a known change in body weight to its energy equivalent. Of course, this assumes that the chemical composition of the body remains uniform during growth;

if it does not, computation requires much more detailed sampling of the population and greater numbers of animals. However, knowing the energy equivalent of body product gives no indication of the energy cost of growth, which will depend on the actual chemical composition of the body—i.e. the relative amounts of different sorts of body products synthesized. It is quite possible to determine these products, although the small sizes of many animals in wild populations may impose serious practical difficulties. However, if the approximate energy equivalents of body products plus the work of growth and the basal metabolism are known the cost of activity can be easily found by subtracting these from the net energy. This point will be referred to in Chapter 4.

It must be remembered that net energy is all that the animal has to utilize. Since activity requirements compete with body products (including growth) for this energy a range of growths on a single value for net energy is theoretically possible. That fish seem to be among the relatively few organisms demonstrating wide intra-specific differences in growth rate in wild populations is therefore very significant (see Chapters 4 to 9).

Needham (1964) listed other useful terms, relating to the energetics of food, maintenance and growth. *Crude gross efficiency* is the percentage of ingested energy stored as body fabric. *Maintenance* is the energy required to keep the animal alive at constant weight (it includes the operating costs of the organs plus the cost of their repair). *Net efficiency* is the energy added to the body as a percentage of the total energy available minus that used for maintenance. A further definition, due to Rubner, is the *true net efficiency,* the energy that goes into new material as a percentage of its own value plus the work of growth.

The procedure for determining total income capitalization and expenditure of food and energy is often very complicated. In a particular instance it may require the use of such a simple direct method as weighing (together with measurement of oxygen uptake and carbon dioxide output under various conditions, assessment of food intake and a knowledge of the chemical composition of the food) the differing growth of the various organs and data on their chemical compositions which may change as the animal becomes bigger or older. Nearly all the classic work along these lines has been done on man and his animals—domestic and laboratory mammals and birds (Brody, 1945; Kleiber, 1961; Mitchell, 1962-64). Food of selected composition can be fed to captive animals under required conditions of work or rest, and elaborate means for monitoring food eaten, excretion and elimination rates, respiratory requirements etc. are available (see Kleiber, 1961; Watt, 1966). Any proposal to use such methods and concepts in the study of wild animal populations, however, raises a series of problems, some of which are considered in Chapters 4 to 9.

In returning to the question of determining energy utilization by indirect means consider Table 2.2, which deals with heat

Table 2.2
(After Kleiber, 1961)

Calculation of animal heat production by indirect calorimetry.
Example: Holstein cow, wt 970 lb = 440 kg, fourth day of fast

From Carbon and Nitrogen Balance	
N lost in urine (measured)	53 g
Protein catabolized (53 x 5·88)[a]	311 g
C in catabolized protein (311 x 0·52)	162 g
C from protein in urea (53 x 0·43)	23 g
C from protein in respiration (162 – 23)	139 g
C lost as respiratory CO_2 (1275 litres x 0·532)[b]	697 g
C from catabolized fat (697 – 139)	558 g
Fat catabolized (558/0·765)	729 g
Heat production from protein (311 x 4·8)	1490 kcal
Heat production from fat (729 x 9·5)	6890 kcal
Heat production per day (from N and C balances)	8380 kcal
From Oxygen Consumption	
O_2 consumed per day (measured and reduced to 0°C, 760 mm, dry)	1730 litres
Heat production per day from O_2 consumption (1730 x 4·7)	8132 kcal

[a] Assumed that meat protein has 17% N.
[b] 1 litre CO_2 = 1/22·4 mol CO_2 = 12/22·4 = 0·532 g C.

production of a fasting Holstein cow. The first part of the table shows how to calculate heat production from the carbon and nitrogen balance of the animal. The nitrogen lost in the urine must first be obtained. From that figure, because "meat" protein (the mixed protein of muscle and other tissues) contains about 17% nitrogen, the total amount of protein catabolized can be derived. Since carbon represents about 52% of the dry weight of "meat" protein, by multiplying the protein value by 0·52 we can obtain the amount of carbon in catabolized protein. The carbon lost in urea may also be obtained by multiplying the nitrogen in the urine by the appropriate factor. All carbon save that in the urea must have been expired as CO_2, unless there has been a change in the acid-base balance of the blood. To separate carbon derived from fat from that derived from protein catabolism (already known by computation), subtract the latter value from the total weight of carbon expired. To convert the weight of fat and protein catabolized to calories, multiply them by their heat equivalents, and thus by summation the animal's heat production is derived.

These computations begin with only two measured quantities: nitrogen lost in urine and carbon lost as respiratory CO_2. The remaining steps in the calculation are logical inferences based on knowledge of conversion factors, such as that for nitrogen to "meat" protein or heat production from fat and protein, the values of which were empirically determined in earlier studies.

The calculation of heat production from oxygen consumed is also given in the Table 2.2 for comparison. It has been found empirically that 4·7 Cal/l. is an appropriate conversion factor to calculate heat production from oxygen consumed assuming the metabolism of a mixture of fats and proteins in a fasting mammal. This is not an entirely satisfactory check on the method of indirect computation from carbon and nitrogen balance. Both methods are indirect and subject to errors inherent in the approximate empirical conversion factors they employ. But direct calorimetry, although technically more satisfying, is a very unwieldy method for larger animals, demanding their complete confinement for a prolonged period under abnormal conditions. Indirect calorimetry is a valuable technique for ecologists, because from only a few sets of measurements they can often make numerous important inferences concerning metabolism.

The above calculations do not, of course, refer to energy which the body would be storing in the case of a growing animal; they only show the results, in the form of liberated heat, of the operations of the kinetic energy systems of the body—and those in the fasting animal. In the growing animal much energy goes into the *work of growth* (see Fig. 2.8) and into general maintenance (i.e. compensation for wear and tear and "minimal" metabolism). The stored products of the body, such as keratin, bone, scales, fat depots etc. which have a negligible or very slow turnover rate, represent potential energy that is more or less fixed and unavailable. Hair is for example unrecoverable to the body once formed, though the hard substance of bone does have a slow turnover. Certain fat depots are readily available for metabolism while others are almost completely unavailable. Glycogen stored in the liver represents potential energy which can readily become kinetic. Muscles and liver, apart from the continuous flow of energy they release in their own maintenance metabolism and the larger amounts that accompany their role in the more vivid demands of activity, may waste away if food is scarce or activity excessive; and, whether it occurs in young, rapidly growing or in old mature animals, such wastage represents degrowth and therefore a loss of potential energy.

The above calculations (Table 2.2) deal specifically with a fasting homeotherm. The periodical effects of the specific calorigenic action of foods are thus avoided and the animal is metabolizing only fat and a little protein. A more complex model would be required for an animal in a less artificial or restricted state; see Chapters 4, 5, 8 and 10.

Table 2.3
(After Kleiber, 1961)

Estimation of heat production from fat and carbohydrate catabolism

	Heat Production in kcal		
RQ $\left(\frac{\text{mols } CO_2}{\text{mols } O_2}\right)$	Per Mol CO_2 $\left(\frac{87}{RQ} + 27\right)$	Per Mol O_2 $(87 + 27\,RQ)$	Per Litre O_2 $\left(\frac{87 + 27\,RQ}{22{\cdot}4}\right)$
0·707	150	106	4·7
0·75	143	107	4·8
0·80	136	109	4·9
0·85	129	110	4·9
0·90	124	111	5·0
0·95	119	113	5·0
1·00	114	114	5·1

Kleiber provided many additional examples of values useful for indirect estimates of catabolic heat energy equivalents. Table 2.3, based on one of his tables, illustrates the heat production that may be assumed in homeotherms according to their respiratory quotients (ratio of CO_2 released to O_2 consumed) when they are catabolizing mixtures of fats and carbohydrates. The heat production values are calculated from the following relationships:

$$\Delta Q/\text{mol } CO_2 = \frac{87}{RQ} + 27$$

$$\Delta Q/\text{mol } O_2 = 87 + 27\, RQ$$

where ΔQ = heat production in Cal/mol for a range of RQ's from 0·707 (pure fat catabolism) to 1·00 (pure hexose catabolism). The heat production for 0_2 consumed converted to Cal/l. is obtained by dividing the values for Cal/mol O_2 by 22·4. Table 2.4, based on another of Kleiber's tables, demonstrates the same sort of relationship as Table 2.3. Here, heat production is based on catabolism of mixtures of fats and proteins (RQ range 0·707 to 0·83), and

$$\Delta Q/\text{mol } CO_2 = \frac{95{\cdot}4}{RQ} + 15$$

and

$$\Delta Q/\text{mol } O_2 = 95{\cdot}4 + 15\, RQ$$

where ΔQ = heat production in Cal/mol. As before, conversion of 0_2 consumed to Cal/l. is obtained by dividing by 22·4. Indirect calculations of heat production on two different sorts of diets can therefore be made merely from a knowledge of CO_2 output and O_2 uptake (since from these values RQ's can also be computed).

Table 2.4
(After Kleiber, 1961)

Estimation of heat production from RQ of fasting animals catabolizing only fat and protein

Catabolizing	RQ	Heat Production in kcal: Per Mol CO_2 Produced $(95{\cdot}4/RQ + 15)$	Per Mol O_2 Consumed $(95{\cdot}4 + 15\,RQ)$	Per Litre O_2 Consumed $\left(\frac{95{\cdot}4 + 15\,RQ}{22{\cdot}4}\right)$
Fat	0·707	150	106	4·7
Mixture	0·72	147	106	4·7
Mixture	0·74	144	106	4·7
Mixture	0·76	141	107	4·7
Mixture	0·78	137	107	4·8
Mixture	0·80	134	107	4·8
Protein	0·83	130	108	4·8

To apply physiological methods in ecological research is, as already pointed out, often very difficult. A physiological investigator may impose a restriction of activity on his experimental animals totally unlike anything they could encounter in nature, making his results impossible to apply to the ecologist's frame of reference. The investigator who wants physiological information for ecological purposes may therefore have to make appropriate adaptations of the physiologist's findings or carry out the necessary investigations himself.

Specifically, since net energy, as defined above, is literally all the animal has to carry out its vital activities, the amount that finds its way into growth will be determined by the needs of maintenance and activity. So, to know how useful food is to an animal, it may be necessary to use not only standard indirect calorimetry but also to find ways to account for activity in wild populations and compute its cost.

A scheme to depict heat production in a homeotherm is relevant here (Fig. 2.9) and indicates the things that augment heat production as temperature falls in fed and unfed animals. Taking first a fed animal (heavy line of main curve), over the body temperature range 13° to 37° C, minimum heat production remains the same. The 13° C temperature line is *Tci*, the "critical

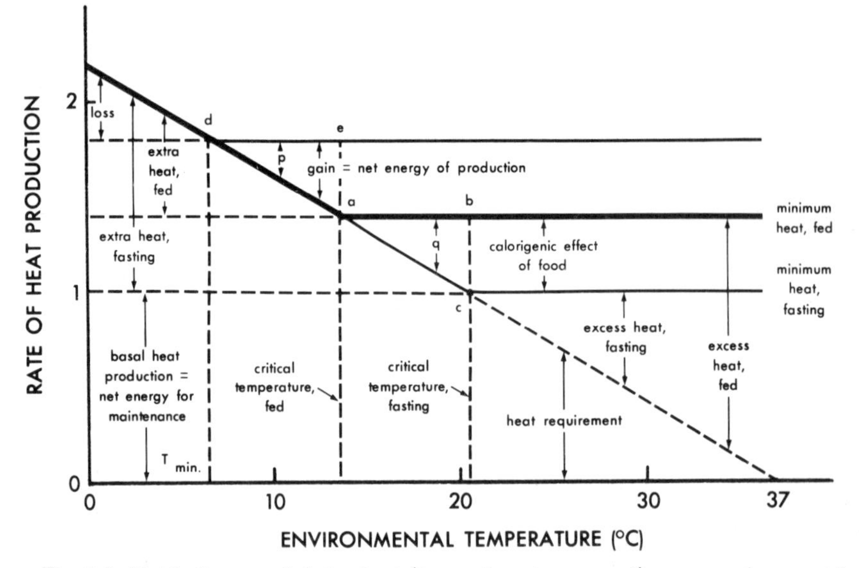

Fig. 2.9. Metabolic rate of fed animal (homeothermic mammal) versus environmental temperature. (After Kleiber, 1961.)

temperature" of the fed animal; at lower temperatures the calorigenic effect of the food is absorbed by the rising heat production which accompanies the maintenance of body temperature. Above this temperature the effect of the calorigenic action increases to a maximum which coincides with *Tco* the critical temperature for an unfed animal (which experiences no calorigenic action), which of course indicates the lowest temperature at which *it* may display minimal heat production. While the figure is almost self-explanatory the facts implicit in the extended sloping line are of special relevance, because both fed and unfed animals may have their heat budgets explained by reference to this slope. In the case of the fed animal, the heat requirement to keep body temperature constant (without any excess of energy for production) decreases as the environmental temperature is increased from the critical temperature *Tci* to 37°C. Correspondingly, the excess heat increases. As the environmental temperature falls below *Tci* the heat production in the fed animal rises (while body temperature remains constant). This increase involves the use of all the heat produced by specific calorigenic action of food and progressive loss of energy which would, at temperatures higher than *Tci*, go into body product, so that growth is no longer possible. Eventually the energy stored in the remainder of the body is also called on, so that wasting occurs.

The same arguments apply to a fasting animal, except that the critical temperature *Tco* is higher because no food is being consumed so no heat from calorigenic action can be contributed to the rising need for heat. Since there is no food there can be no body product and, indeed, the fasting animal will utilize the fabric of its body to augment its heat output when the environmental temperature falls below *Tco*. By comparison the fed animal will not actually waste until the environmental temperature reaches *Tmin*.

This scheme is somewhat idealized and simplified. For instance, it implies that the activity needs of the animal are to be met over the whole temperature range *Tmin* to 37°C from a constant quantity of energy available for growth, the work of growth maintenance and activity. Yet if the animal's activity becomes large this supply will be insufficient to meet its needs. Wastage of body fabric will follow.

Theoretically the animal, fed or unfed, will counter decreasing temperature by increasing inroads on stored energy. However homeotherms from polar and high mountain regions, therefore adapted to cold, may remain in positive heat balance and thus active, and may grow, at environmental temperatures far below freezing because of their efficient insulation.

The sorts of relationships in this scheme are therefore of most direct use in understanding the physiological-thermal relations of domestic animals living at intermediate temperature ranges. Within these ranges certain mathematical relationships probably hold fairly well. For instance, Kleiber gives the following mathematics for Fig. 2.9:

From the triangle *abc* covering the changing part of the calorigenic effect, it can be seen that this effect (q) at any temperature (T) between the critical temperature of the fed (Tci) and the fasting (Tco) animal can be calculated from the proportion

$$\frac{q}{T - Tci} = \frac{q \max}{Tco - Tci}$$

$$q = q \max \frac{T - Tci}{Tco - Tci}$$

and we can calculate the rate of animal production, ρ, between the critical temperature of the fed animal and the minimal temperature for production. From the triangle *aed* it follows that

$$\frac{\rho}{T - Tmin} = \frac{\rho}{Tci - Tmin}$$

where

ρ = rate of production of body substance, milk etc.
ρ max = maximum production rate with the ratio considered
T = environmental temperature
$Tmin$ = minimum temperature at which production is possible
Tci = critical temperature of the fed animal.

This scheme, though it is not the only one which could be offered for all homeotherms, gives certain working principles needed to understand production for growth in them and suitably modified it could be adopted for most homeotherms. It also gives a theoretical framework for a whole group of inferences about the metabolic balance of animals.

Kleiber (1961) has clarified another, often confusing, matter regarding homeotherms. Since smaller mammals have a higher metabolic rate per unit weight it appears that they are more "wasteful" converters of food to flesh than bigger mammals. Kleiber's example, in which one steer (1300 lb) is compared with 300 rabbits (same total weight), shows that though the total heat loss per day of the rabbits is four times that of the steer (as is their relative food consumption), in consuming 1 ton of hay, they each make total weight gains of 240 lb. Yet even though they "waste" so much more heat per day, the rabbits gain a total of 8 lb per day against 2 lb per day for the steer. In strictly economic terms the rabbits are in many ways far the better proposition.

Such considerations will have very important future applications in studies of fish growth and production.

Fish, although vertebrates, are neither homeotherms nor behavioural thermoregulators like many reptiles. Many species of fish seek preferred temperature ranges in their natural environment which certainly may influence their whole behavioural life and their growth rate, through optimizing metabolism. But most fish do not appear to be storers of heat. Indeed their aquatic way of life would make it very difficult to conserve heat against or dissipate it into the external environment, unless they possessed insulation and thermoregulatory abilities to match aquatic mammals. Water itself has a high heat storage capacity and a high specific heat, and fish body substance has properties similar to this medium, whereas terrestrial animals have the favourable heat storage properties of water within their own bodies and ready means for heat dissipation into an atmosphere of relatively low heat storage capacity. All that terrestrial animals need is effective insulation against the passive continuous drainage of heat to the exterior and a good means of reducing or by-passing this insulation for rapid heat dissipation where required. This, birds and mammals have evolved. The ectothermic reptiles lack good insulation and the independent thermostatic control over basal metabolism to allow them to generate enough heat for a high body temperature; but once having got warm by behavioural adjustment they can maintain body temperature, even in cool air, by appropriate additional behaviour, far better than a fish could in cool water.

Fish, then, are poikilotherms, in so far as that term still applies, as are the invertebrate remainder of the animal kingdom. A scheme suitable to illustrate growth in homeotherms will be totally inadequate for fish and other poikilotherms. And although the principles of energy utilization and storage in Kleiber's diagram are generally applicable to homeotherms, the differences between the physiology and living conditions of fish and homeotherms mean that the details of a comparable scheme for fish would be considerably different.

III. PHYSIOLOGICAL AND CELLULAR ASPECTS

A. Cells and Growth

The concern here will be with cell growth—multiplication and hypertrophy—rather than cell or tissue differentiation. Needham (1964) observed that many instances of differentiation are results of changes in the relative numbers of different types of cell, i.e. that they are growth phenomena. He considered confining the term "differentiation" to changes in the body's organization rather than "mere addition of units", though such a distinction must be somewhat arbitrary. Needham also proposed discarding the recurrent argument of whether differentiation and growth are mutually exclusive—with differentiation sometimes inhibiting growth—by supposing that they proceed in successive cycles.

It is possible to catalogue the cell populations of the vertebrate body, so that all cells are placed in one of three categories: *renewing, expanding* or *static* cell populations. The following list is based on Goss (1964).

Renewing cell populations
Epidermis (and derivatives)
Endodermal epithelium
Endometrium
Transitional epithelium
Gonadal germinative cells
Haemopoietic tissues
Skeletal tissues (in part)

Expanding cell populations
Liver
Kidney
Exocrine glands
Endocrine glands
Lens
Connective tissues proper
Skeletal tissues (in part)

Static cell populations
Striated muscle
Neurones
Neural retina

The expanding cell populations do not, in the full-grown adult vertebrate, expand further. Space available within the body and the proximity of other organs and tissues may play some part in controlling any expanding tendency. However, *in vivo* experiments do reveal that a repeatedly ablated liver (for example) has the power to regenerate repeatedly almost its original bulk (Goss, 1964, 1967) and it does this by increasing its cell number. Furthermore, as Goss reported, static tissues are only a particular case of expanding tissues in which mitotic growth is confined to early development. Of course, even post-mitotic tissues may grow by cellular hypertrophy (e.g. muscles) and the distinction between expanding and static cell populations must be essentially arbitrary.

Bertalanffy (1960) indicated some of the rates at which cells may be formed daily in various tissues, his data coming mainly from mammals. Tissues such as nerve, adrenal medulla, muscle (equivalent to Goss's static cell populations) receive, in the normal state in the adult animal, no new cell additions each day. These are fixed, post-mitotic, highly differentiated or specialized cells, rarely dividing and then only in response to an acute need for hyperplasia or regeneration. Other tissues (approximating to Goss's expanding cell populations) such as adrenal cortex, liver, kidney and thyroid, are also post-mitotic and do not normally divide, except where regeneration or repair are called for. A small number of cells (less than 1%) may be formed each day, not for tissue renewal but merely to compensate for the losses of wear and tear. In vertebrates, it is claimed, the only way the kidney can grow is by hypertrophy of nephrons—that is by hypertrophy and hyperplasia of tubule cells either during normal growth or through increased functional demand. Enlargement of nephrons beyond a certain limit leads to renal insufficiency (Goss, 1967).

Another group of cells (Goss's renewing populations) consists of tissues such as intestinal epithelium, stomach lining cells, epidermis, bladder and tracheal epithelia, lymph nodes, etc. with "a relatively high mitotic rate in the adult organisms which is balanced by a corresponding loss of cells, so that the total cell number remains unchanged and neither positive nor negative growth takes place." Such populations have daily cell addition rates ranging from 70% (intestinal epithelium) to 1.5% (bladder epithelium) with corresponding renewal times ranging from 1½ to 67 days.

Obviously in early, sometimes embryonic, growth, all cell populations will be multiplying rapidly. Those that become most highly differentiated or functionally specialized lose most of their mitotic ability—e.g. muscles and nervous systems. It used to be thought that in vertebrate muscles the final number of cells was attained early in life and remained constant thereafter, further growth in size of muscles resulting merely from increase in size of cells (Goss, 1967). Cheek and his associates (Cheek, 1968a, b, c; Cheek *et al.*, 1968) have, however, reported that in boys muscle cell number increases from 5 to 16 years with a peak rate of increase at 10 to 11 years. Muscle cell number doubles from then until 16 years. Cell size, also, shows a sustained linear increase from infancy through adolescence to adulthood. In girls, the number of cells increases more rapidly but reaches its limit at 10½ years. The increase in number of muscle cells in post-natal life is 10-fold in girls, 14-fold in boys. Similar values and trends have been found for rats (Cheek *et al.*, 1968).

The tissues that retain their early power of cell division and which therefore have considerable capacity for cell replacement are less highly specialized or differentiated. These are the tissues which are much exposed to wear and tear in daily life. In the absence of all damage such capacity might lead to unacceptable tissue hyperplasia.

Bertalanffy (1960) has emphasized that protein turnover is only loosely tied to tissue renewal rate. It is high in intestinal mucosa, for instance, which correlates with the high renewal rate, but is also high in liver which is usually mitotic only in regeneration; here the correlation is with the high functional activity of the liver tissue. Nervous tissue, with cell constancy, also has a high rate of protein turnover.

Protein synthesis and cell growth appear to be under the control of cell nuclear DNA. That is, their control is genetically determined in the first instance. When cells divide, cytoplasmic changes, including increase in RNA, appear to precede nuclear changes. And the nuclear DNA content does not change even when degenerative changes in cell cytoplasm are quite advanced. Apparently, too, the nuclear volume bears a fairly constant relation to the 2/3 power of cell volume, a simple case of allometry which holds in haploidy, diploidy, tetraploidy and even in mutants. It is perhaps significant that in certain pathological conditions there is an upset in the usual constancy of this "nucleocytoplasmic" ratio (Bertalanffy, 1960).

It is probably profitless to expect any very general answer to the problem of the sources of energy required for mitosis. Bertalanffy (1960) stated (pp. 159-160) that "Animal ova which represent the simplest conditions, show all possible differences as to whether or not mitosis can be carried out under anaerobic conditions, depending on the species concerned and its particular adaptive conditions." However, exponential growth phases do seem linked with anaerobic glycolysis.

Bertalanffy (1960) gave a simple model to account for different growth patterns displayed by different cell types in culture. In pointing out that rod-like cells (many bacteria) show a simple exponential growth curve, whereas spherical cells (yeast) have a sigmoid curve of growth, he postulated that because rod cells grow only in length, the surface available for nutrient and respiratory exchange remains directly proportional to volume, while as a sphere grows the "surface-volume ratio is shifted in disfavour of surface."

For rod-like cells the following equations will apply:

$$dv/dt = \eta v - xv = cv$$

$$v = v_0 e^{ct}$$

$$l = l_0 e^{ct/3}$$

(v = cell volume; η, x = constants of anabolism and catabolism, respectively; c = growth constant; l = length; v_0, l_0 = initial volume, length). A constant specific growth rate and a simple exponential growth curve will represent the situation.

In spherical cells specific growth rate decreases and

$$dv/dt = \eta v^{2/3} - xv. \tag{2.4}$$

For linear growth (in radius = r) the decaying exponential

$$r = r^* - (r^* - r_0)e^{-kt}$$

and for volume growth a sigmoidal curve

$$v = [\sqrt[3]{v^*} - (\sqrt[3]{v^*} - \sqrt[3]{v_0})e^{-kt}]^3$$

(r^*, v^* = final radius, volume; r_0, v_0 = initial radius volume; $k = x/3$).

Furthermore, the equation (2.4) above, if it applies to spherical cells, helps to account for the principle of constancy of cell size: "If a growth equation of (this kind) applies, the growing system is equifinal, i.e. it will attain the same final size independent of initial size."

Why cells divide at a particular point and time in their life history is obscure. In one sense, genetical factors must provide the controls. But from the time of D'Arcy Thompson (1942) it has been apparent that physical forces in the cell itself, which are intimately related to the stresses and structural relations of its size and shape, feature importantly.

Since the days of early microscopy cells have seemed to represent the building blocks of the body. Though not all cells in all tissues exist as discrete units, they do so frequently enough to claim that they constitute one of the great natural categories of organization of living material. Cells of many metazoan tissues will live apart from the body and tissues that form their natural milieu. Some cells can be cultured indefinitely, though in most cases they appear to lose their original character after a time *in vitro,* as Hayflick (1968) stressed. Sometimes they will also form structures in culture that resemble the tissues of which they form a part in the living animal or new structures unlike anything seen in the body.

Since so many tissues in culture apparently have the power to continue to grow by division while space and nutrients are not limiting it appears that those two factors may act as controls of tissue growth in the body.

Certain investigators, notably W. S. Bullough and his colleagues, consider that self-secreted, tissue-specific substances termed chalones act, in higher concentrations, on various tissues so as to limit proliferative growth through depression of cellular mitotic activity (Bullough and Laurence, 1967). In this view, only those cells in a "dying state" are unlikely to revert, under certain conditions, from an essentially functional to an essentially mitotic condition. The view also seems to imply (Bullough and Laurence, 1967) that such post-mitotic tissues as most nerve cells and striated muscles may be in a "dying state"—at least from this point of view of control of cell proliferation. It appears that neurones and striated muscle cells are lost progressively during a lifetime (Curtis, 1963; Shock, 1962; Bullough and Laurence, 1967). Bullough (1971) has also proposed, partly on the basis of experiments on mice, that adrenaline and a glucocorticoid—both secretions of the adrenals—act in conjunction with chalones, which inhibit mitosis, to inhibit cell ageing. Mice, consuming a restricted diet that was nevertheless adequate to maintain full health, showed enlarged adrenal glands, together with a marked reduction in mitosis in epidermis and sebaceous glands. Since the skin did not thicken nor sebaceous glands increase in size it was concluded that a reduction in cell proliferation brought about by this hunger stress was balanced by a lowering of cell death. Bullough (1971) also wants to extend this concept to help explain some aspects of ageing in animals.

Teir *et al.* (1967) described experiments on rats which suggested that, as tissues undergo normal necrosis at the end of their lives, they release a substance that stimulates proliferation among remaining cells. Many other relevant studies on control of cellular aspects of growth have been described in the illuminating series of essays edited by Teir and Rytömaa (1967).

Cell biologists have found that in cultures containing more than one cell type, normal cells "tend to sort out into groups of pure type" (Needham, 1964). Cancer cells do not possess this ability. Affinities between some different types of cells are also evident in cultures, for instance ectoderm and endoderm (there are junctions between cells of these types at the extremities of fore- and hind-gut). These properties of attraction and repulsion, the absence of either capacity in cancer cells and the ability of nerve cells in culture to connect different sorts of cell clumps suggest that similar forces may operate in the developing animal as tissues differentiate and in preserving the integrity of tissues once formed. It is, in fact, neither tissues nor cells that produce the growth pattern of a metazoan. It is rather as if the organism started its life with a small collection of cell cultures and proceeded to direct, modify and manipulate the growth, proliferation, differentiation and life expectancy of those cultures. In doing so the body, in its centrally directive role, may to some extent intricately blend or organize these cultures at different places and times. The various cultures themselves may frequently become the environments of other cultures and *vice versa.* During the life of the organism each of the different cell groups may display versatility in repairing damage, responding to different stimuli and by dividing into several sorts of tissue.

Fell and Mellanby (1953) and Fell (1962) described how skin explants of week-old chick embryo when cultured in media, containing vitamin A excess, secreted mucin instead of keratin and formed active cilia on the epidermis. This gives remarkable *in vitro* evidence of the unexpected extent of structural possibilities that cells may reveal under unusual conditions.

It has also been demonstrated *in vitro* that once cells have been embryologically determined as to their type they will adhere to this type by forming characteristic structures or tissues unless seriously interfered with. Thus, when cells of one tissue are cultured, in suitable media, with those of other tissues they frequently form aggregates that resemble histologically the tissue from which they are derived. For instance, mixed cell suspensions of limb-bud and mesonephros of chick produced recognizable regions in the culture of cartilage, skin-like structures and nephric structures. Though the obvious inference here would be that cells had retained their original identities following dispersal in culture and eventually recombined into structures characteristic of their type, it was just possible that transformation or metaplasia of cells had occurred in culture. A way of resolving this difficulty was to make mixed aggregates of cells of chick liver and mouse and chick pre-cartilage, when the hepatic tissue forming *in vitro* was of chick cells alone (microscopically recognizable), whereas cartilage masses that formed were chimaeras. No transformation of cell types was evident. However, because only cells which were already embryologically determined as to type were used there remains a chance that cells from earlier embryos would not display so "determined" an outcome in mixed culture (Moscona, 1965).

Kleiber (1961) returned to the perennial problem of whether differences in metabolism between whole animals can be explained as an inevitable outcome of differences in the metabolic rates of their tissues. He concluded that the correct explanation must include a role for central regulators; in other words, it is not merely additive effects of tissues that produce a given oxygen demand, but the organism as a whole, as conditioned by its own central thermoregulatory systems. But it is cells which actually use oxygen and produce heat, so that Kleiber can also agree with the view that "central regulators operate by influencing cellular factors". Kleiber adds that although there are important genetically determined differences between animals (mammals), this knowledge "... does not help in the interpretation of the fact that the metabolic rate per unit weight of a 5·3 kg rabbit is practically the same as that of a 6·6 kg dog, but differs significantly from that of a 1·5 kg rabbit, as in turn the metabolic rate per unit weight of a 2·5 kg dog differs significantly from that of the 6·6 dog." Nor does it help in interpreting the fact that while an unborn rat foetus has a metabolic rate *in vitro* similar to the rate per unit body weight of the intact mother, it has a much higher rate after it is born or when 12 days old.

Kleiber's conclusion was that "cellular metabolic rate is adapted to the condition of the animal as a whole ..." The cells, in other words, have considerable plasticity of metabolic behaviour. It is the role of the whole organized self-regulating system of the intact animal to determine the metabolic rate, per unit of weight, at which the cells operate.

Earlier we considered growth in terms of curves and bio-energetics of whole organisms: their overall growth progression or the major energy transactions they carried on with the environment.

Above we have very briefly considered cells. The next section treats the essential link between these two categories: the physiological integration and regulation of growth. Cells are the basic units of growth, and the major result of their operations is overt increase (or decrease) in size of the individual organism or its maintenance in a steady state.

B. Regulation and Integration of Growth

Processes such as multiplication or replacement of tissues, which proceed against a background of degrowth of other tissues or which amount to mere compensation for the loss of tissues through damage, wear and tear or death of senescent cells, are basically similar to the processes that result in the growth of an entire organism. In the adult animal these processes may only at certain critical times outweigh cell losses or degrowth processes, though they may be going on at all times. Certain marine invertebrates demonstrate that during the extreme remodelling of metamorphosis, it is apparently biologically economical for the parts of the adult body to originate from small groups of cells in the larval body, which is subsequently discarded (Needham, 1964). This is a considerable difference to the holmetabolous insects in which the adult structures arise from imaginal discs present in the larval body. Though this larval body is entirely disorganized during metamorphic remodelling, nearly all the original larval material is reused by the adult as it forms itself.

The time of appearance of organs in the animal and their subsequent rates and patterns of growth and their final bulk and function depend on cell growth, but, more importantly, on phenomena such as competition for the materials needed for growth among the various organs and upon their final role and position in the body. Patterns of use and disuse of organs in the growing body will also alter their size—possibly at the expense of other organs. The increased use of muscles leads, as is well known, to their increase in cross-section and therefore contractile strength. This may be at the expense of fat in other organs, but that depends on the total food supply available, as the intake may simply be adjusted to the augmented metabolic demands of the more active and enlarged muscles.

Various organs and tissues display different growth abilities. Just as it is resistant to the effects of malnutrition in the mature animal, the nervous system seems to grow more satisfactorily than in most other tissues in animals on a poor nutritional plane. The same is true of the ovary (Needham, 1964).

Much of the organizing ability of the body during growth goes into the synthesis of the skeleton. In the vertebrates, proteins form the progenitor of the skeleton in its correct orientation and only as it is completed are the calcareous materials laid down in this fibrous matrix.

The intermediary metabolism of amino acids is one of the most important sets of processes during growth. All higher animals, especially vertebrates, require for protein synthesis essential amino acids which they cannot synthesize themselves and without a supply of which they begin to suffer from deficiency diseases. The 10 essential amino acids for vertebrates (including fish, according to Love, 1970) are the following: arginine, histidine, isoleucine, leucine, lysine, methionine, phenylalanine, threonine, tryptophan and valine. The following (again including fish) are amino acids that can be left out of the food without harm: alanine, aspartic acid, cysteine, glutamic acid, glycine, hydroxyproline, proline, serine and tyrosine.

Not long ago it was thought that only the small percentage of ingested amino acids required to compensate for loss and wear found their way into the "fabric" proteins of the animal body. The remainder were thought to be used as a source of energy. Radioactive tracer studies have now shown that perhaps 50% of ingested amino acids pass into the body fabric causing displacement and eventual excretion of an equivalent amount of amino acids from body proteins. If amino acid intake is increased displacement rate is stepped up accordingly. In true growth (neoformation), it is thought that the newly ingested amino acids are utilized. Thus (Needham, 1964):

Inflow (of amino acids from food) is comprised of neoformation + repair + variable exchange fraction;

Outflow (of amino acids) is comprised of damage + wear and tear + variable exchange fraction.

True growth accounts for a major fraction of turnover in actively growing animals (turnover being the daily flux of amino acids caused by the displacement from the body tissues by those newly ingested). DNA turnover is strongly related to growth, being 20 times higher in newborn rats than in adult ones. Furthermore, uptake of amino acids into tissues bears a strong positive correlation to the different tissue growth rates and productive activities, being very high in intestinal mucosa and low in muscles, brain and collagen of connective tissue.

Anterior pituitary growth hormone (APGH) helps growth, evidently by promoting protein storage rather than neoformation, so that amino acids "which previously were merely displacing their identity in the fabric are now taking part in neoformation." It is possible that this may represent a general principle of dietary economy affecting, also, inorganic materials in the food and in the body (Needham, 1964).

C. Vitamins

Many vitamins are concerned in growth (Needham, 1964, gives a list and refers to their probable roles; see also Wagner and Folkers, 1964; Karlson, 1963). Vitamin D is vital for the normal growth of the skeleton of vertebrates (lack in mammals produces rickets), its serious deficiency over prolonged periods leading to permanent, crippling bone deformity.

Vitamin E (α-tocopherol) may be synergistic with Vitamin A in growth promotion, being specifically needed for normal growth of reproductive organs and of the growth of mammalian foetuses. That Vitamins A and C are deeply involved in preserving the normal structural relations of tissues as they develop has been known for a long time (White, Handler and Smith, 1964). Excellent supplementary data on their effects have become available in recent years through studies on the *in vitro* culture of tissue and organ explants (summarized by Fell and Rinaldini, 1965). Vitamin A, at fairly critical levels, is needed for the integrity of developing epithelial and skeletal structures and their subsequent maintenance. Less is known of the precise role of Vitamin C, but its presence in appropriate concentration is apparently a condition of adequate maintenance and rapid production of collagen in tissues of mesenchymal origin (Gould, 1961; Robertson, 1961). Vitamin E is also related in some fashion to maintenance aspects of growth—the offsetting of wear and tear. Muscular dystrophy may result from its lack.

Vitamin K_1 promotes growth in a general way, while many or all of the B complex of vitamins—usually acting synergistically among themselves—are concerned in the maintenance functions of growth. Generalized dystrophy may result from a failure to ingest enough of the B group of vitamins.

D. Hormonal Influences

Biologists well know that the action of blood-borne hormones on target organs and the delicately balanced feedback mechanisms involved in their secretion are one of the primary means for the coordination and timing of events concerned with the body's economy. Moreover, through neuro-secretion, nerves may release hormones or similar substances which represent in many instances a more direct mediation between nerves and systemic effects, than where a nerve affects an endocrine gland which subsequently releases a hormone.

Students of growth cannot neglect hormones, because they affect growth—its extent, duration and intensity. The most significant of the hormones for growth in vertebrates is anterior pituitary growth hormone (APGH), the following account of which is taken largely from the review by Knobil and Hotchkiss (1964). In man and mammals, in which classical studies of its role have been carried out, APGH is a prime controller of growth. Insufficient secretion of growth hormone in man can, according to the time and duration of its insufficiency, produce several kinds of dwarfism, stunting or wasting of varying degrees of severity; its over-production can likewise lead to more than one kind of growth stimulus resulting in giantism in the young or acromegaly in adults.

As yet the precise character of the growth hormone molecule remains uncertain though there appears to be a degree of group specificity. Pituitary extracts from birds show only slight effects in the rat, while reptile and amphibian extracts are much more effective—yet extracts from fish are without effect. Cattle pituitary extracts can promote growth in fish but primate extracts are less active, although both act similarly on rats. There certainly does not seem to be a particular phyletic ordering in these effects. Even among primates, despite immunological cross-reaction between antisera of different members, there is evidence that the hormone is not identical in all.

Non-metabolizable α-aminobutyric acid labelled with C^{14} has been used to discover that APGH promotes amino acid accumulation in tissues, which is apparently how it mediates growth. Added to diaphragm muscle from hypophysectomized rats, bovine growth hormone increases cellular uptake *in vitro* of a wide range of amino acids. Those that are rapidly metabolized on entering the cell show least tendency to accumulate, as is probably to be expected.

The presence of pancreatic hormone, insulin, is necessary for the full anabolic effect of APGH; both hormones promote amino acid accumulation, though some evidence suggests that APGH also directly stimulates protein synthesis. APGH augments fat oxidation as indicated by lowered fat storage, increased ketogenesis and a reduced respiratory quotient; it also shows a hypoglycaemic action.

The thyroid influences growth as well. Thyroxine (tetraiodothyronine), the main hormone of the thyroid glands, controls metabolism in homeotherms and its secretion rate is, in turn, controlled by the activity of pituitary thyrotropic hormone. It has important growth effects in young animals, being not only directly stimulatory but also governing metamorphosis in amphibians and fish. Stimulation of growth of nervous system and bone are among major effects of thyroxine; its effects on bone are brought about synergistically with APGH. In homeotherms thyroxine overproduction can greatly increase metabolic rate, while thyroid insufficiency leads to an abnormally low metabolic rate. The level of thyroxine production could affect growth through changes in the anabolism-catabolism ratio at all periods of life and also according to whether the processes of growth are adding to body mass or acting in a regenerative or compensatory manner.

Lasnitzki (1965) reviewed the evidence of the effects of various hormones and hormone-producing tissues on growth and maintenance of tissues in culture media. In general it appears that hormones are capable of affecting tissues directly, often in ways that appear to parallel their effects in the whole animal. The growth influence of insulin is especially striking in that it promotes proliferation of cells in culture and glycogen and lipid synthesis. RNA synthesis is also increased. Growth is promoted as an accompaniment of rising glucose utilization, but it is not sure whether cell proliferation demands more glucose or whether a rising glucose uptake results in mitosis. The effects of insulin on bone growth are very manifest in cultures of embryonic chick bones. The typical effects are the production of a short-shafted bone, with enlarged ends due to the overgrowth of epiphyses. The effects of APGH in culture are more equivocal and seem largely to demonstrate that insulin can, as well as combining with APGH, sometimes be antagonistic in its effects.

Adrenocortical hormones inhibited growth of cultures of rabbit fibroplasts (Gagianut, 1951), while chick-embryo heart, spleen and spinal cord treated with cortisone acetate or other cortical substances were also inhibited (Ruskin *et al.,* 1951). Lasnitzki (1965) described other studies that led to similar results. In this connection, it is interesting to recall the postulates of Bullough (1971) concerning a relationship between glucocorticoids, adrenalin and chalone production, already referred to above. However, Lasnitzki (1965) has noted that it "is probable that the growth inhibition and interference with mitosis seen in so many different cells are due to a non-specific toxic effect . . . not related to their physiological function."

The promise that cell and tissue culture holds for future growth analysis may be summarized as follows:

(i) The direct effects of hormone-secreting tissues on cells may be assessed directly and simply by microscopic examination.

(ii) Since living cells are the tangible product of these experiments they can be evaluated by weighing, counting, chemical analysis or by an appropriate bioassay or physiological study.

(iii) The advantage of working *in vitro* and with small quantities of tissue is that experiments may be readily replicated and that the delays required for an experimental result are short in comparison to the time usually needed for whole animal experiments.

3 | Growth of Fish in Populations

I. THE PROBLEM AND ITS SCOPE

The ecologist's study of growth will begin and end with growing animals in populations. It is, after all, the patterns of growth of animals in populations and the effects of these patterns on the dynamics of the populations that gives the ecologist his justification for an interest in growth. Subsequent analysis of growth, including laboratory study and theoretical interpretation, may lead to improved understanding of growth processes, in turn resulting in better interpretation and prediction of ecological consequences of growth for the population. But determining how rapidly fish grow in populations can be a demanding problem, leading to fallacious estimates unless the methodology is sound. This chapter considers the determination of the growth of fish in wild populations and attempts to state the essentials of a methodology which will not be misleading.

To assess growth in a population of very simple size and age structure (e.g. a number of young fish newly released in a pond that can be fully drained) is an easy matter which fish culturists the world over handle as a matter of everyday routine. Most wild fish populations are, however, represented by an assortment of individuals of both sexes, ranging in some instances from newly hatched larval fish to adults many years old. Furthermore, both the age and size structures of the population (i.e. the frequency of different ages and sizes of individuals) may be very complex, resulting as they do from a highly dynamic series of factors including food, space, mortality etc. Nor is it usually possible to sample most populations with much frequency. It is therefore important to develop techniques for determining growth from the characteristics displayed by members of the population at the time of their capture, assuming that an effective method of capture is available to allow for the collection of a large number of animals that are not seriously biased as to size or age. Frequently, the age and size structures of populations may be used to obtain important clues about growth, sometimes even definitive evidence. However, it is obvious that if the age of individuals of different sizes can be readily determined—assuming age-for-size is not too variable—we have a powerful tool for population age estimation. And if a sufficiently sensitive and reliable method for determining the age-for-size of particular individuals exists the extension of this technique is clearly of extreme value to the fish population analyst.

Basic procedures for age and growth determination in fish have been well stated by Tesch (1968). His is a detailed and fully documented account and contains many points of valuable technical advice. The present chapter considers in some detail a few selected studies on the assumption that if general principles are to be found they will underly *all* such studies and that the best way to reveal them is by examination in depth of a limited number of problems. It is, however, appropriate to note that much of the basic study and application of the methodology of age and growth determination in fish populations was carried out on marine fish of great commercial importance. Hodgson (1957) gives a very readable, non-technical account of this for the herring fisheries of the Northern Hemisphere (see also Lea, 1910, 1913, 1938; Lee, 1920). Thompson, H. (1923, 1929) gave a notable application of the scale reading method to haddock biology.

II. METHODS OF DETERMINING AGE

Le Cren (1947) listed clearly and comprehensively the requirements for deciding the validity of supposed age marks on scales, otoliths, operculars and other bones of fish. These requirements are given below, redrafted to allow for rather different emphasis of certain points and with comments on the usefulness and limitations of the various methods. As Le Cren pointed out, he did not originate the criteria he used which were based on Graham (1929) and Van Oosten (1928, 1941).

A. Length-Frequency Curves

Populations of fish that reproduce seasonally are characterized by regular influxes of new recruits so that if adequately sampled they usually reveal a size structure featuring a train of wave-like modes (length-frequency polygons), especially in the first few years of life, which indicates the presence of several age groups (Fig. 3.1). Should individuals from such age groups bear on body structures (such as scales, otoliths, operculars, vertebrae or fin spines) marks corresponding to the order of the fish in the size structure of the

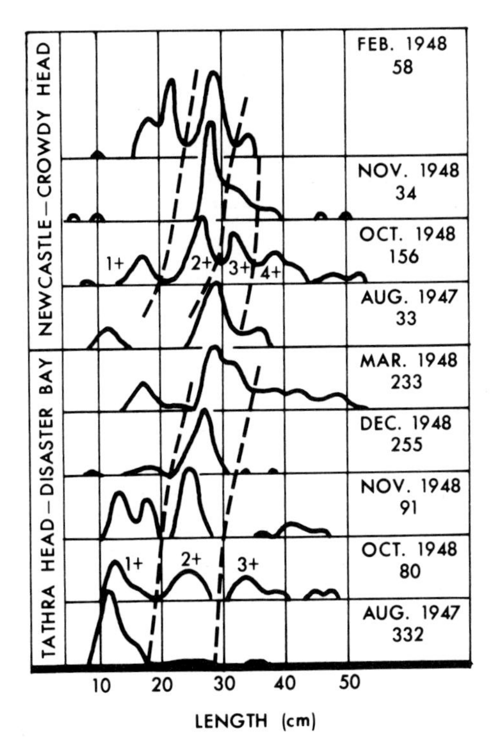

Fig. 3.1. Length frequency curves of flathead *Neoplatycephalus macrodon* obtained by experimental trawling; samples from the southern and northern areas only, arranged according to months irrespective of year. (Simplified after Fairbridge, 1951.)

population, these may be used as indicators of age. Further, if the modal lengths should correspond to mean lengths-for-age as derived from a valid method of back-calculation (see later) then, in some instances, this may be taken as additional evidence of the validity of marks on scales or other structures as age indicators. This is the Petersen method, named after its originator.

Comment: The fact that changes in the environment of the population may produce considerable changes in the density of its various component age classes can cause different age classes to have widely differing growth records. For this reason, back-calculated growth from the scales of an old year-class may bear little relation to the actual growth of younger year-classes and hence to the modal groups in the population. Furthermore, for most teleost populations, the utility of this method declines with age. As fish become older, individual growth rate differences cause the ranges of their lengths-for-age to increase more or less constantly. The differences between size groups, often so clear among younger year-classes in the population, thus become blurred. Such modal extinction is accelerated by mortality which diminishes the absolute numbers in all age groups throughout the life of the fish. Lastly, the method is relatively useless in the tropics "where young fish may enter the population throughout the year and . . . where life cycles are often short" (Tesch, 1968). Olsen (1954) gave an interesting account of his attempt to apply this method to the ageing of the school shark *Galeorhinus australis* which, as an elasmobranch, lacks hard bony parts of the sort normally available for age determination in teleosts.

B. Abundant or Scarce Year-classes

Recruitment is always variable, but sometimes a combination of favourable conditions results in an unusually large recruitment, i.e. a new year-class of exceptional strength is added to the population. The members of such a year-class must, like all the others, progress in mean body size as it ages, usually suffering no more or less than the normal mortality rate and therefore persisting as a notably strong size mode. If it is possible to sample regularly a population containing such a peculiar size mode the scales or other suitable structures of individuals from the mode may be examined. Addition of a new mark to these structures each year would indicate the validity of such marks as age indicators, especially if it can be established that each new mark was put down at the same time of the year (see below). It should be noted that unfavourable conditions at spawning or during egg–larval development can sometimes produce a particularly small subsequent year-class. This may be as obvious, as it progresses as a size mode through the population, as would be an unusually large year-class.
Comment: Even when the population lacks a year-class that is outstanding because of its unusual abundance or scarcity, it is frequently possible to note the progression of the ordinary modes through the population with time. This method is often used in fishery science, in conjunction with growth rate studies based on use

of scales, clipped fins, tags or other marks; it may be used by itself in cases where scales or other skeletal elements are too difficult to interpret or where it is impracticable to mark or tag, usefully, large numbers of fish. Its utility decreases with the older year classes, for the reasons given above.

C. Examination of the Growing Edge of the Scale or other Bony Structure

If scales or other bony structures are sampled frequently enough from a growing population, the supposed year mark, if reliable, will appear at the growing edge during a relatively short time each year. As the scale or other structure continues to grow the margin surrounding the annulus will increase in width until the next annulus makes its appearance.

Comment: This method of establishing the validity of the year marks is widely used (Table 3.1). If, however, the marks at the growing edge of scales etc. are found to make their appearance over an extensive part of the year (e.g. six months), or even randomly in time, they will be useless. This method is used mostly by those investigating large commercial fisheries because, like the Petersen, it requires samples of considerable size collected frequently for at least a year.

D. Marking Individual Fish

Fish can be marked (e.g. by fin-clipping, tatooing, staining with dyes etc.) before release at an early age and before they have any scales, at least before the scales etc. (as revealed by examination) have any obvious marks on them; they can be marked at any known age (or size), tagged with a numbered tag of metal or plastic or injected with tetracycline which produces a mark in the structure of the otoliths before release; scale samples are collected at the same time if required. At recapture the number of supposed age marks on the scales etc. should coincide either with the known age or with the number of years the fish have been at liberty.

Comment: This method is the most satisfactory because of the certainty of the knowledge that it affords. The main drawback is technical. It has been difficult to mark very young fish in an entirely harmless fashion–though this difficulty can now be overcome sometimes. Bigger fish will usually bear a tag quite well, though

Table 3.1
(After Thomson, 1957)

Seasonal growth of II+ yellow-eye mullet calculated from marginal increments of 1025 fish. All localities Western Australia. Measurements in centimetres.

No. of Annuli	Jan.	Feb.	Mar.	Apr.	May	June	July	Aug.	Sept.	Oct.	Nov.	Dec.
1	4·7	7·0	7·4	8·6	8·5	8·5	8·4	8·3	8·8	0·6		
2	4·7	6·4	6·8	5·9	7·0	7·2		7·2	7·2 (1·1)[a]	1·0	2·0	3·8
3	3·6		4·3	4·8	5·5	5·9			5·7 (0·5)[a]	0·8		
4		2·7	3·1		4·1	4·2		4·1	4·1			
5			2·0									
No. of fish	116	84	210	188	386	177	78	101	200	115	20	84

[a] The figures in parentheses indicate the increments at a later date in the month.

occasionally it has been claimed that growth has suffered either through interference with feeding or by irritation caused by the tag (Holt, 1962). The recruitment of fry into many wild populations is enormous and to put a marking experiment on a sound basis, great numbers of young fish often need to be released. The same is true of older year-classes, which are easier to mark, but may be difficult to capture and keep in health in the requisite numbers, before and after they are marked. Of course, even relatively few marked fish, if there is a chance that an appreciable proportion of them may be recoverable, can yield direct information of great value in a particular situation (Olsen, 1954). Clearly, where the chance of recovery is high, fewer marks may yield adequate data, especially where the species has a high growth rate so that relevant information can be obtained quickly.

A very full review of the methods used to mark fish in Australia has been given by Thomson (1962). Stott (1968) has also provided a comprehensive account of this subject.

The marking method becomes almost the only approach to age determination in species, such as elasmobranchs, in which hard parts are unavailable and (for older fish) length-frequency distributions are suspect (Olsen, 1954).

E. Walford Graphs

Walford (1946) found a useful method for determining the annual length increments to be expected at any initial length. The method enables the investigator to use fragmentary length data from a population sample consisting of tagged individuals of a wide variety of sizes and ages, so that by plotting length at age $t + 1$ against length at age t a curve may be obtained which, if each year's increment is a constant fraction of the preceding year's, may be represented as a straight line. The plotted line lies above the diagonal (45° slope). but in many instances converges towards it with age. In some cases a line parallel to the diagonal may be obtained, which means that there is a uniform absolute increase in length with age.

Comment: The advantage of a Walford graph is that one may derive an estimate of average growth in the population from the meagre data obtained by recapturing relatively few tagged fish and, if one can determine a single length-for-age, the remaining lengths-for-age can be readily determined from the regression line. However, the method assumes a steady state population in which growth rate remains largely constant over many years.

F. Rearing and Holding Experiments

Fish ponds or large fish tanks offer means for testing the validity of marks on bony structures. If the fish are of known age or size when placed in the ponds it is easy to recapture them later and examine scales or other structures for presence of the number of marks to be expected, if these are annual occurrences.
Comment: This method is naturally used more frequently on fish of inland waters, because of the relative difficulty of holding marine fish in small enclosures for long periods.

III. CASE STUDY AND COMMENTARY: THE WINDERMERE PERCH POPULATION

Our major case study of age determination comes from the work of Le Cren (1947) on perch (*Perca fluviatilis*) in Lake Windermere. Until his investigation it had been usual for workers on this species to employ scales, opercular bones or both. Le Cren's decision in favour of operculars followed the discovery about 1940 that in polarized light, rings on operculars were easier to see; they also gave indications of being more reliable than those on scales. Le Cren set out to determine whether opercular rings were true annuli.† He discusses the appearance of the operculars, the method of removing them, their orientation for viewing in polarized light and so on; but of more relevance here is his description of the "broad, opaque zones that correspond to the rapid growth of summer, and each of (which) gradually fades into a narrow transparent winter zone, which ends relatively abruptly, with a sharp line of discontinuity between it and the next summer zone."

The means that Le Cren adopted in testing the validity of age determination from operculars involved the three methods as given earlier.

Under the first he obtained small fish from Windermere over a five-year period at various times of the year (Fig. 3.2) and from different parts of the lake. Slight local growth rate differences tended "to increase the variance of size in each age group and to make the

† "Annuli" are annual year-marks and therefore reliable age indicators.

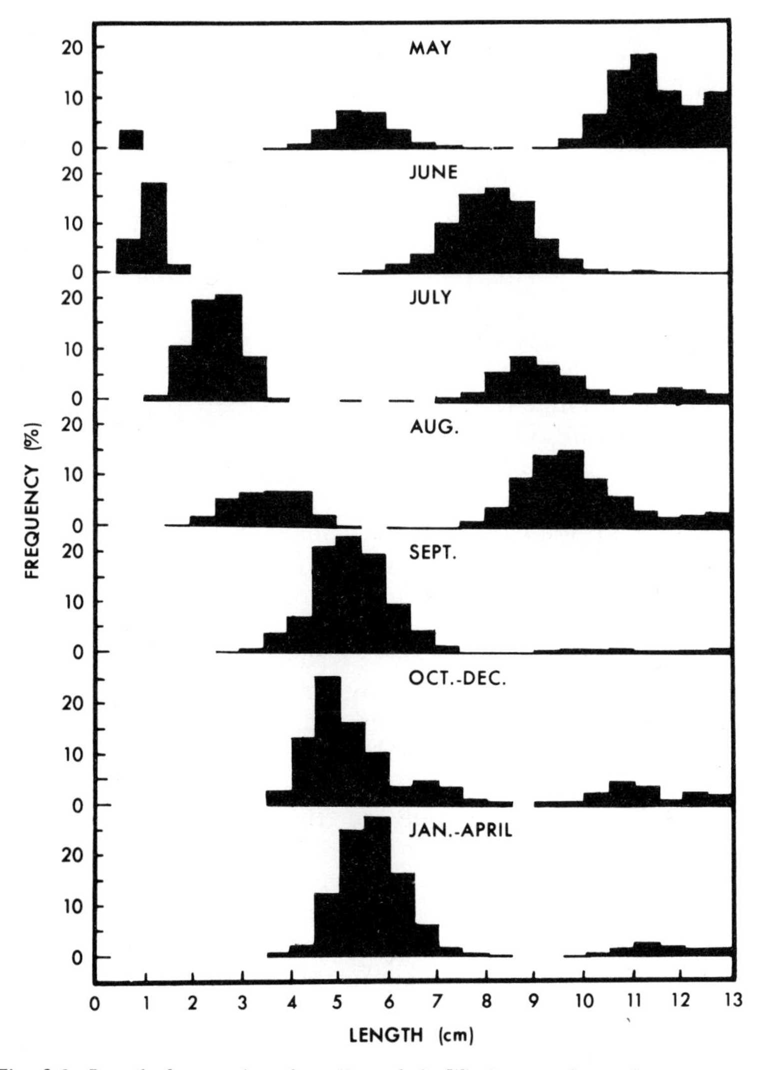

Fig. 3.2. Length frequencies of small perch in Windermere for each summer month, autumn and spring. (After Le Cren, 1947.)

length-frequency distribution flatter than normal." The figure shows clearly, however, that among these small perch (less than 12·5 cm) there were O, I and II year-classes.† May-June was the time of appearance of the O group, immediately after hatching, which then grew to between 5 and 6 cm by the end of the summer, when growth ceased each year; the O group of one year would become the I group of the following year to reach a modal size of about 11 cm by summer's end. As II group fish in the next succeeding year they would exceed 13 cm.

On examination according to the criteria given earlier Le Cren found that the operculars of O group individuals showed no ring before the fish had gone through the first winter of life. By the following summer they had become I group individuals (i.e. showed one ring surrounded by a summer growth band the width of which depended on the time that had elapsed since the preceding winter). Similarly the third size mode (II group) showed in summer two winter rings. The length modes, then, coincided with the numbers of rings (0, 1, 2) on the operculars of the fish comprising the modes.

Figure 3.3 shows data from a sample of fish caught in spring (May), plotted into length-frequency polygons, with the corresponding age analysis indicated by the operculars. The modes (two each for male and female) correspond with fish bearing two opercular rings and three or four opercular rings. The latter signifies that the second size mode is already composed of two more or less completely overlapping year-classes; i.e. after only three years of growth we see the confusing overlap between age groups mentioned earlier.

Table 3.2 gives an example of the progression through the population of a somewhat older age group (described earlier): the IV group of 1944. This was a relatively abundant size group, though it should be noted that O and I group fish were not usually big enough to be taken by the method of capture used (fishing in the same place and month in successive years). This progression of IV group to V group was determined by examination of operculars.

The relatively few anomalies encountered among the operculars Le Cren examined were mostly due to unusual crowding together of the rings as a result of slow growth.

There is a huge specialist literature on the analysis of marks on

† O, I, II . . . etc. is a standard notation for the number of years of life that fish have survived. An alternative notation is 0+, 1+, 2+ . . . etc. In each system the O, or 0+, age group signifies that fish have not yet reached the end of their first year of life.

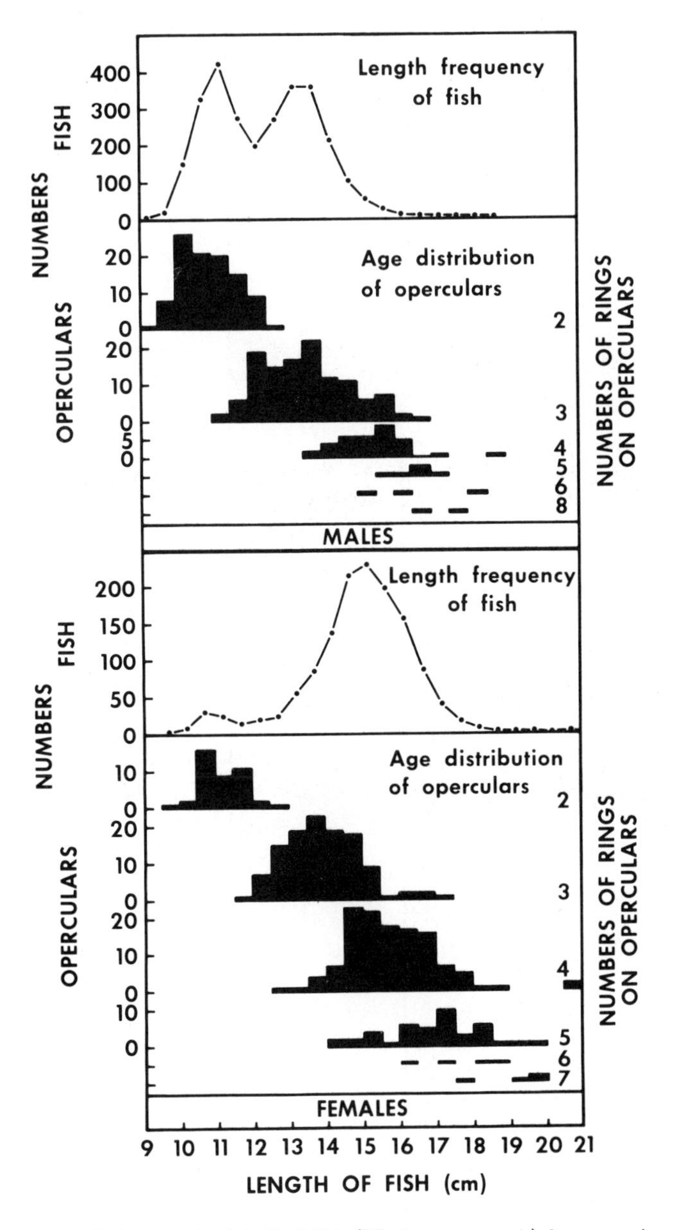

Fig. 3.3. Length frequencies (a) of all fish (Windermere perch) from a series of traps in 1944 and (b) of smaller samples classified into age groups according to the number of winter rings on their operculars. (Le Cren, 1947.)

Table 3.2.
(After Le Cren, 1947)

Percentage age distribution (age groups I–XI) from opercular readings of two samples of perch angled in October at one station in 1944 and 1945. (All females, the one male caught in 1944 omitted)

Year	Total Number of Fish	I	II	III	IV	V	VI	VII	VIII	IX	X	XI
1944	124	0·8	0·0	12·1	57·2	21·0	6·5	0·0	0·8	0·0	0·8	0·8
1945	95	0·0	0·0	0·0	14·7	71·6	11·6	2·1	0·0	0·0	0·0	0·0

operculars, otoliths, scales etc., much of it in journals devoted to hydriobiology, fishery science, ecology and ichthyology. Some of the more prominent of these journals are listed in the Appendix; Tesch (1968) provides much guidance. Of this voluminous literature surprisingly little has had, over the years, that definitive quality which renders the use of age determination methods as unequivocal as one would wish. And much of the literature describes not the testing of the methods on new material, but merely the application to new fishery problems of methods already worked out—which is natural enough and to be expected as the outcome of any successfully developed methodology.

A few other examples of papers that cover with particular success this question of age determination from scales and other bony structures are mentioned at the end of this chapter. Whichever actual element of the growing fish is employed for age determination, the criteria that must be met are those already discussed. If such criteria cannot be applied the investigator usually must concentrate on obtaining evidence from marked fish or attempt to analyse the population into size modes. The procedure of using scales etc. for age determination is, after all, a matter of convenience. If it is obviously inadequate it is certainly not convenient.

One of the main pitfalls of the approach to age and growth determination via population structure includes the effect of gear selectivity,† which frequently means that fish below a certain size are not represented—at least not in proportion to their real abundance—in the population sample. Gear selectivity may therefore suggest a population structure that lacks one or more of the age-size classes of the actual population. In some instances fish become too *large* to be taken by nets, traps or lines used for sampling the remainder of the population so that they, too, may fail to appear as part of the population structure revealed by the capture methods employed. Even a method such as electrofishing usually fails to sample very small (and sometimes very large) specimens in a population, unless special adaptations of technique are adopted.

These, and related procedural difficulties, are more easily countered in inland waters where the whole life history of fish populations is often fairly well known, including movements, behaviour and general habitat of fish of various ages and sizes. For example, it is sometimes possible, even in large lakes, to sample a

† Clearly, the structure of fish populations, both as to age and size, revealed by a sample of a commercially exploited population, is subject to the effects of gear selectivity. For nearly all commercial fishing gear is selective—some of it highly so.

complete range of year-classes effectively enough to gain a full picture of population age and size structure.

Nowadays a variety of techniques is available for sampling fish populations, ranging from intensive use of traps and nets (with mesh sizes which vary from that of plankton netting to outsize graball net), lines, poisons and electrofishing gear; nearly all have undergone considerable development over the last 25 years. As acceptable alternatives to each other, or used in combination, these various techniques provide fairly effective means for capturing adequate, and sometimes proportionally representative, numbers of the various year-classes (Lagler, 1968). The major problems are still associated with knowing definitely where the various age groups are distributed, because many species show widely differing modes of behaviour and habitat with age and size. The would-be sampler of fish populations therefore cannot afford to skimp on his natural history.

Parrish (1958) points out that: "The great merit of Petersen's method is that it is a very economic method to use since only length measurements have to be taken, and the laborious and costly process of age interpretation from scales, otoliths etc. is avoided." If size classes were always as clear cut as they can be, the method would certainly be the obvious one to use if all that was required was a first approximation of the mean growth rate of the population. Frequently, however, the method of catching the fish is suitable for taking representative samples only within a certain size range; or the different age-size classes have different behaviour patterns, so that they may appear at one time and disappear at another for reasons unconnected with mortality rates (e.g. Fig. 3.4).

IV. BACK-CALCULATION OF GROWTH

At best, the Petersen method can give only an approximation of the age and growth of fish in a population. In one version of the Petersen method, recruitment of an unusually large or small year-class permits its rate of progression as a recognizable mode through the population over a number of years to be readily followed. Yet, the unusual strength of such a "marked" year-class may, because the competitive intensities within it are unusual, cause the fish comprising it to grow at an unusual rate. And these effects may not be confined just to the one unusual year-class but may change the growth pattern of the entire population. The method of direct marking is in some ways the best, because we can be certain of the precise increase in length of a

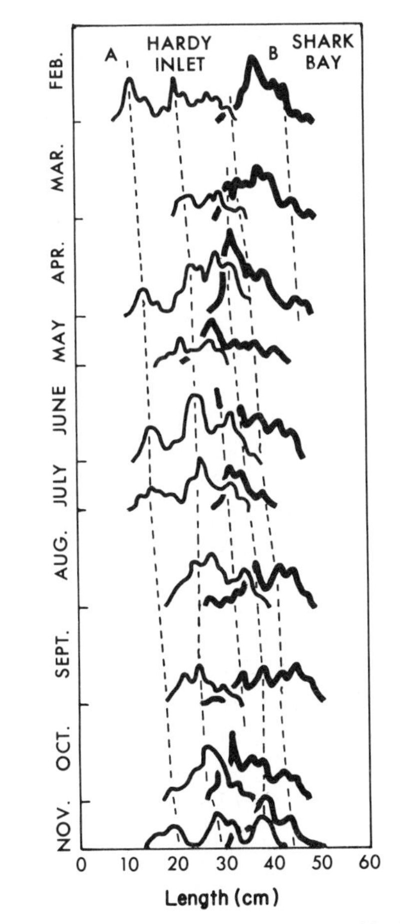

Fig. 3.4. Length frequency distributions of sea mullet *Mugil dobula* in two Western Australian environments, showing modal progression. (After Thomson, 1951.)

given individual over a particular period. But tagged specimens usually require a long time to provide worthwhile information, the return on expense and labour often being very low.

An ecologist who can discover a valid method of age determination from permanent marks laid down in hard parts of the body can age population samples of any size routinely, rapidly and accurately; Hodgson (1957) has described how such a method was applied very effectively to an analysis of the herring of various ages taken by East Anglian fleets. Moreover, if the increase in size of the hard structure bears either a direct linear relationship to the increase in length of the animal, or one which can be accounted for in simple mathematical terms, then the growth history of every individual that it is possible to age can also be computed. With such a research tool, interpretation of populations can assume new dimensions. It sometimes becomes possible to discern and evaluate key events in the lives of animals which, though they occurred many years ago, are reflected in their pattern of growth. The long-term effects of environmental changes can sometimes be quantified in the form of growth rate change if one can compute past growth. The effects of excessive mortality or particular year-class strength are often seen in changing growth rate. From the practical standpoint effects of changing fishing or predatory pressures can be inferred from data on back-calculated growth.

Two examples from the copious literature demonstrate the problems of relating growth of fish to growth of opercular bone and growth of scale, and the procedures adopted by different workers in solving them. These are simply typical examples of sound methodology. Many studies of fish growth merely adopt the methods of earlier investigations and possess no intrinsically valuable features. Many more are unsatisfactory in that the methods of age determination have been inadequately worked out or else the material available has proved after full investigation to be unusable.

If the growth of a scale or other bony element is directly proportional to the growth of the fish then

$$Fx = Fy \frac{Bx}{By} \quad (3.1)$$

where Bx = length (or diameter, etc.) of the scale or bone at a particular age, Fx = the corresponding length of the fish, By = final length of the scale and Fy = final length of the fish.

Le Cren (1947) found for perch that opercular length plotted against fish length gave a slight curve. The logarithm of opercular length minus the logarithm of fish length against fish length gave a marked curve, indicating true curvilinearity in the relationship, not ascribable merely to greater variance of opercular length in the larger specimens. Le Cren assumed a heterogenic (allometric) growth relationship and fitted a regression line to the logarithms of mean fish lengths on the logarithms of mean opercular lengths, for fish from 8 to 21 cm. From the regression coefficients he derived an empirical formula for allometric growth ($Y = bx^k$; Huxley, 1932), of the form

$$\log \text{fish length} = 1{\cdot}357 + 0{\cdot}9202 \log \text{opercular length}$$

or

$$F = 22{\cdot}76\, B^{0{\cdot}9202}$$

Here F = fish length, B = opercular length and 0·9202 signifies the ratio between the logarithnic growth rates of the fish and opercular (i.e. slope of regression line). This was termed the "growth ratio", after Huxley (1932).

The direct proportion formula given above (3.1) would then become

$$Fx = Fy \frac{Bx^{0 \cdot 9202}}{By^{0 \cdot 9202}}$$

also expressed as

$$\log Fx = \log Fy + 0 \cdot 9202 \, (\log Bx - \log By).$$

Le Cren constructed a chart of the curve of this equation from which, by use of a straight edge, the corrected length-for-age could be read off the y (fish length) axis.

The growth ratio for perch was 0·9202 (nearly unity); in case it might be thought that this could be ignored in practice, Le Cren compared the use of this growth ratio, with the method of direct proportion, to back-calculate growth from a perch 20 cm long. If a length of 10 cm be assumed by direct proportion, the growth ratio would give 10·6 cm (i.e. a 6% adjustment needed). However, at a length of 2 cm by direct proportion the corrected length would be 2·5 cm (25% difference).

Le Cren sounded a cautionary note in pointing out that:

It is difficult to compare samples of the Windermere perch population taken at different times. First, samples of perch tend to vary in mean size and other characters with the region of the lake from which they come. Thus for comparative purposes samples should come from the same locality. Secondly, most methods of fishing tend to be selective in one way or another so that many catches are not good random samples of the population. Thirdly, over the period of investigation the population has been subjected to a very heavy and selective mortality by the trap fishery, so that in back-calculating one or more years, not the whole original population is being sampled, but only that part of it which has survived selection by the fishery. As well as this mortality due to the fishery there is a natural mortality which may well tend to eliminate certain sized individuals. Thus it cannot be expected that samples from the survivors of one or more seasons of fishing and natural mortality will give an accurate picture of the original population.

This summarizes the main difficulties of adequate sampling in fish population studies concerned with growth and is a valuable warning for those involved in such work. It appears, from Le Cren's subsequent analysis of Windermere perch, that the effects he referred to were not, in this instance, sufficient to vitiate the tests he applied.

Le Cren applied further tests to his back-calculations. He caught perch in the autumn when growth had ceased and recorded their lengths and ages (I, II, III and IV groups). In succeeding years he checked these lengths-for-age by back-calculation from operculars of specimens of these same year-classes grown older and larger. His results are shown in Tables 3.3 and 3.4. None of the differences between direct measurements and back-calculations of length-for-age are so large that they are unlikely to be due to chance.

O group fish were also netted in spring, just after winter (when growth should be stationary), in four successive years (1940-43). In 1944 three collections of older fish yielded data from back-calculation which indicate lengths for O group fish at their first winter significantly different from the direct measurements, in a few instances. The differences were so small as to be of little importance, but Le Cren suggested that they might be due to a higher mortality directed towards the smaller fish in the first year or so of life (Table 3.5).

The same type of analysis was extended to many hundreds of additional perch ranging from 10 to 20 cm in length. There were some minor difficulties largely attributable to sampling problems. For instance, among perch of age groups I to IV caught in traps in the spring of 1943 "while in this case there are differences between the means from I year's samples and those back-calculated from the next year's samples (1944), these differences are of the same magnitude as those between results obtained in the same year from two different samples."

Finally, recapture of 15 tagged perch also gave good general confirmation of the method of growth analysis.

This investigation by Le Cren must be regarded as a success. It began with careful examination of growth rings, went on to consider in critical detail the relation between growth of opercular and growth of fish (for which an effective formula was determined) and, armed with an apparently sound means for correcting back-calculated growth, provided a successful series of tests of the accuracy of this method. Le Cren was aided by working on a lake fish population. The formidable problems of sampling effectively in the sea (e.g. herring; Hodgson, 1957) are minimized in many inland waters. Moreover Windermere is limnologically one of the best-known water bodies in the world and studies of the perch populations were begun in the 1930's.

Our next example is concerned with the need to establish precisely the relationship between growth of fish and growth of scales.

When Nicholls (1957) began his investigation of the Tasmanian

Table 3.3
(After Le Cren, 1947)

Female fish caught in pilot traps. Numbers in samples, mean lengths and standard deviations of four pairs of samples. The first in each case is obtained from measurements of actual fish; the second by back-calculations from the same year-class in a later year. The age group and the year caught of the sample measured or used for back calculation are given in each case, as well as the value of P in a t test applied to the difference between the means of each pair.

Age Group	I	II	II	III	II	IV	III	IV
Year	1942	1943	1942	1943	1942	1944	1943	1944
Mean length (cm)	10·89	11·23	14·24	14·13	14·24	13·87	15·87	15·81
Number	25	23	50	51	50	11	51	11
S.D.	0·554	0·888	0·977	0·849	0·977	0·886	0·905	0·557
P	$0{\cdot}11 < P < 0{\cdot}12$		$0{\cdot}54 < P < 0{\cdot}55$		$0{\cdot}21 < P < 0{\cdot}22$		$0{\cdot}77 < P < 0{\cdot}78$	

Table 3.4
(After Le Cren, 1947)

Similar comparisons to those in Table 3.3 for two age groups of female fish angled in 1944 and 1945

Age Group Year	IV 1944	V 1945	V 1944	VI 1945
Mean length (cm)	17·92	17·90	17·94	17·87
Number	71	68	26	11
S.D.	0·950	0·782	1·119	0·964
P	$0{\cdot}89 < P < 0{\cdot}90$		$0{\cdot}84 < P < 0{\cdot}85$	

Table 3.5
(After Le Cren, 1947)

Numbers in samples, mean lengths and standard deviations of O group fish in the spring of 1940, 1941, 1942 and 1943 obtained from measurements of netted samples, and by back-calculations from three samples of older fish netted in 1944.

Year	Statistic	Actual Measurements from Netted Samples	Back-calculated from Operculars 26 July 1944	25 Aug. 1944	28 Sept. 1944
1940	Number	201	4	3	0
	Mean length (cm)	5·07[a]	6·19[a]	6·05	—
	S.D.	0·65	0·49	1·26	—
1941	Number	129	25	20	10
	Mean length (cm)	6·19	5·96	6·21	6·17
	S.D.	0·81	0·75	0·91	0·91
1942	Number	467	25	9	15
	Mean length (cm)	5·43	5·45	5·67	5·53
	S.D.	0·60	0·59	0·71	0·92
1943	Number	192	62	31	11
	Mean length (cm)	5·46[b]	5·39	5·18[b]	5·57[b]
	S.D.	0·60	0·43	0·70	0·45

[a] For the difference between these means by t test $P < 10^{-5}$.
[b] For the difference between these means by t test $0{\cdot}03 < P < 0{\cdot}04$.

trout fishery he found that, to be sure of data from back-calculation of growth, he had to examine the relation between scale length and fish length. An essentially straightforward problem, the reason for its importance makes this study of interest to fish ecologists.

Trout, along with salmon, were among the first fish whose scales were critically scrutinized in connection with age and growth (see the

Appendix of this chapter). The most important technical difficulties associated with the interpretation of trout scales were worked out long ago in the Northern Hemisphere. But Nicholls had to be sure that the standard of his data would satisfy those working on trout in other countries. As Tasmania is an island roughly the size of Scotland, to which brown trout were introduced in 1864 and rainbow trout in the 1890's (Weatherley and Lake, 1967), it seems not unreasonable to suspect that a certain amount of genetic divergence from the Northern Hemisphere stocks of both species might have occurred over such a period. Either this, or influences of local conditions, might have affected growth in ways not found elsewhere, making it necessary to be sure that scales carried the same sort of information about trout growth in Tasmania as elsewhere. Nicholls never aimed at a fundamental study of the validity of year marks on trout scales. He used the established criteria, for their recognition, to check that they appeared with an annual regularity. Where spurious marks occurred he also used accepted criteria to identify these. Nicholls identified false winter rings by their short duration in the midst of a normal summer growth band. Certain trout from lakes lacked one or two annuli, but these fish were identifiable from the presence of the annuli on the scales of the majority of other trout in the population. Hatchery reared trout showed no evidence of a false first annulus as a result of their artificial living conditions. Jaw-tagged trout did show a mark—presumably indicative of a brief interruption of normal feeding—readily distinguishable from normal annuli.

This disposes of the evaluation of the actual marks on scales which Nicholls examined, their presence and configuration corresponding to what was already known of the characteristics of such marks.

Nicholls determined the scale length-to-fish length relationship for 755 brown trout from the North Esk and St. Patrick's Rivers, captured by electrofishing in December 1954 and January 1955. About five scales from each fish were measured, the fish themselves ranging from 6 to 36 cm in length. Results are shown in Fig. 3.5.

The extrapolation of the regression line passed, as shown, very close to the origin (when $Y = 0$, $X = 0{\cdot}039$), so that for this combined sample of fish there is a direct linear proportionality between scale length and fish length. Nicholls showed, however, that by separating the data into two groups, each comprised of the fish from two separate rivers, it was possible to compute two quite distinct regressions. The equations were quite similar and even though in one case the extrapolated regression line intercepted the

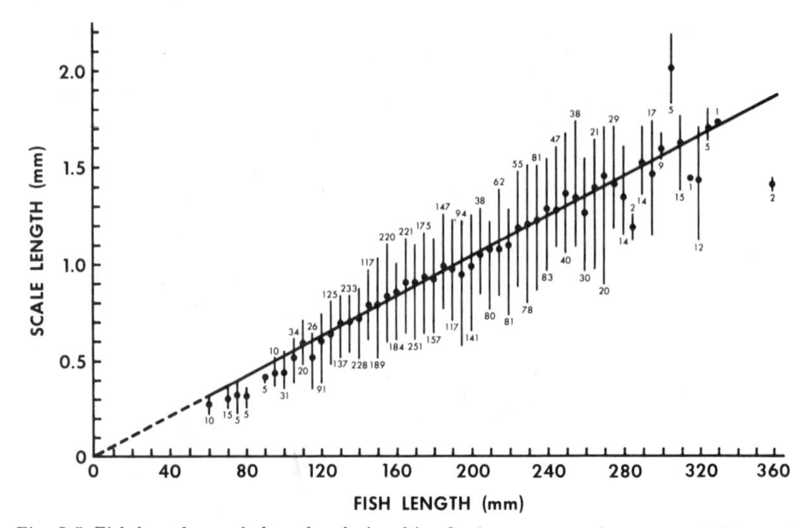

Fig. 3.5 Fish-length : scale-length relationships for brown trout from natural (Tasmanian) sources, ranging from 60 to 360 mm in length. Only the mean values are plotted, but the vertical lines indicate the range of scale length for each fish length and the numbers of observations of each fish length are given. (After Nicholls, 1957.)

x-axis at +1·05 cm and in the other case at −1·24 cm, these differences cancelled out when the two sets of data were combined, the regression line then passing "close to the origin". This is particularly interesting, because both these rivers are similar though quite distinct sections of the one system.

Lee (1912) reported that extrapolated lines of regression of scale length on fish length cut the x-axis at different positive values of x for different species. Nicholls postulated that these differences were due to the small number of values taken by Lee (about 200): "The significance of this would appear to be that regression lines based on small samples are likely to be only approximate." It follows from his data that Lee's claim† that the position at which the x-axis is cut indicates the fish length at which scales first appear is also falsified,

† Jones (1958) has explained this fallacy. The adult scaled fish is covered with scales that overlap—the degree of overlap is constant unless the relation between growth of fish and growth of scale is non-linear. However, fish are born without scales and at the moment of appearance scales are small, discrete platelets. For a short period after formation the scales grow more rapidly than the body of the fish, until they overlap as they will continue to do thereafter. The slope of the regression of scale length on fish length during this period will be much steeper than it is afterwards. Generally this curve is ignored in back-calculation and the main regression of scale on fish is simply extrapolated to cut the x (fish length) axis. If, however, this extrapolated line has an intercept value that departs considerably from zero—as is sometimes the case—that must be allowed for when back-calculating length-for-age.

since he could point to two entirely different intercepts of the x-axis (for closely related monospecific populations), both based on nearly twice the number of fish Lee used. Moreover, one of the intercepts from Nicholls' data was negative, which would, by Lee's reasoning, mean that the fish had scales before they had length.

Nicholls also measured scales of two samples of hatchery-reared brown trout, 251 fish in 1948-49 (3·6-9·9 cm) and 99 fish in 1956 (3·2-8·4 cm), which gave regression lines intercepting the x-axis at 1·71 and 2·03 cm respectively. These values were combined in the regression in Fig. 3.6, which also shows the slope of the line for wild

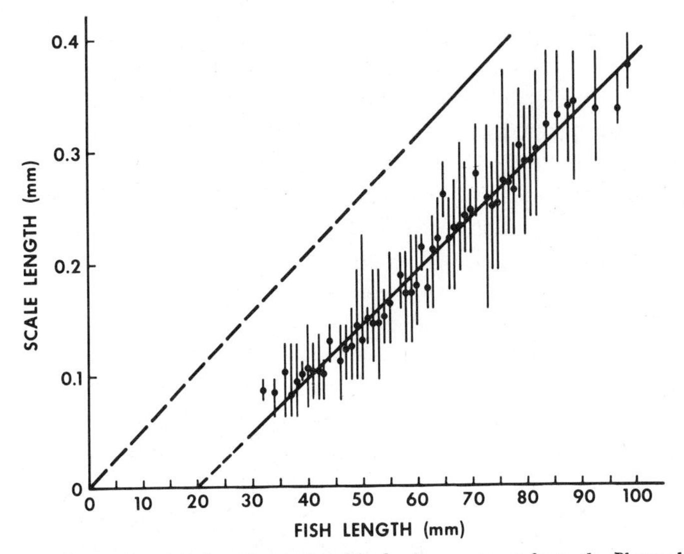

Fig. 3.6. Fish-length : scale-length relationship for brown trout from the Plenty hatchery (Tasmania), ranging from 32 to 99 mm in length. Only mean values are plotted, but the vertical lines indicate the range of scale length for each fish length. The number of observations for each fish length ranged from 3 to 78 with a mean of 9·6 and only 13 sets of data were below this average. The regression line for the data from fish from natural sources given in Fig. 3.5 is included to show similarity of slope.

river fish from Fig. 3.5. Figure 3.6 shows that although the slopes of the lines for hatchery fish and river fish are nearly identical the former had, on the average, smaller scales (which, Nicholls indicated, may be due to a combination of artificial foods and crowded conditions). It also disposes of a possible problem of another sort, as was reported by Blackburn (1950) who claimed for the anchovy *Engraulis australis* different relationships for scale length-to-fish length for fish above and below 6 cm in length. As Nicholls' data from wild fish came from specimens of from 6 to 36 cm he felt it necessary to investigate the relationship for very small trout. The difficulty of getting enough of these from the field caused him to turn to hatchery fish. As shown (Fig. 3.6) the relationship was different, but the slope of the regression line was unchanged and "there is no tendency for the slope of the line to change below 6 cm or at any other point." Nicholls concluded that "for the present purposes it is sufficient to assume that under natural conditions growth of the scale is exactly proportional to growth of the fish throughout life . . ." Nicholls thus provided an example of a study of a relationship that throws new light on an old problem, permitting an investigation of fish growth to proceed in the knowledge that the principal method used is correct in the circumstances of the study.

Although Nicholls' conclusions are appropriate in the case he examined, complex problems of interpretation do arise—Weatherley (1959) caught tench in Tasmania as small as 2·7 cm, yet all had scales. The regression of scale length on fish length for a sample of these fish intercepted the *x* (fish length) axis at +1·7 cm. There might have seemed to be sound reasons for allowing for this value in back-calculations of length-for-age, but Weatherley chose to ignore it, as though the regression had passed through zero on the *x*-axis. This decision was based on the observation that in field collections of tench many fish were 3 cm or less in length in the spring. These were fish that had been spawned the previous summer so that their scales had just one winter check. If the positive (+1·7 cm) interception of the *x*-axis had been taken into consideration in the back-calculation of length, then early growth of the tench would have appeared greater. In particular, Weatherley calculated that L1 would have been about 4 cm—very unlikely judging from actual specimens that had just passed their first winter. On the other hand to ignore the effect of this positive intercept in calculating later lengths-for-age would be relatively unimportant. Weatherley's essential (1959) argument for ignoring it and treating the growth of tench scales relative to length as a case of simple direct proportionality was very similar to that of Nicholls (1957) for trout. However, for a larger sample of tench from another environment with considerably wider scatter of values around the regression line, but lacking data on tench less than 5 cm long, the regression passed through zero on the *x*-axis. It might easily have been assumed that there were differences between these relationships, but for the direct field evidence on the size of young fish already described.

There are, however, other studies which suggest that, at least in some instances, intercepts of the regression of scale versus fish lengths are positive and *real* and must therefore be taken into direct consideration in the computation of length-for-age (Blackburn, 1950; Frost and Kipling, 1959; see also Tesch, 1968).

These studies of the relation between the growth of fish and bony element have important lessons for fish population analysts. As in all science, sensitivity and appropriateness of a method determines the scope and accuracy of the information it will reveal. Le Cren (1947), by finding the curvilinear relationship between growth of opercular and fish, a relationship satisfied by a version of the formula of Huxley for heterogony, was able to improve markedly the analytical strength of the basic procedure. Nicholls' (1957) conclusion about the importance of the point of interception of the x-axis was a negative one; but also valuable, because it disposed of doubts about its possible importance.

In closing this section it is worth repeating Le Cren's (1947) statement:

> Although there is relatively little substantiated direct evidence of the soundness of age and growth determinations . . . the sum of indirect evidence leaves no doubt that in general their use is justified. At the same time the practice of many workers, who have published accounts of age and growth based on the scale method, without attempting to substantiate the validity of its application to the species or problem studied, is to be deprecated.

V. APPENDIX

It is not intended to give even a brief historical survey of the subject of age-determination in fish. However, several publications have been of great importance in the development of methodology. Among these we may cite Lee (1912, 1920), Paget (1920), Graham (1929), Van Oosten (1928, 1941) and Hile (1936).

A series of valuable contributions was published under the convenorship of Parrish (1958), and Ricker's (1958) Handbook is a very valuable work.

A range of journals which print articles emphasizing population and growth of fish include:
Journal du Conseil; The Transactions of the American Fisheries Society; The Journal of the Fisheries Research Board of Canada; The Australian Journal of Marine and Freshwater Research; The Progressive Fish Culturist.

Less frequently work on fish growth appears in:
Ecology; Ecological Monographs; The Journal of Animal Ecology;

The Journal of Applied Ecology; Oikos; Copeia; Hydrobiologia; The Journal of Wildlife Management.

Series of separate publications dealing with fish growth are issued by:

The Food and Agriculture Organization of the United Nations; The United States Department of the Interior (Fish and Wildlife Service); The Fisheries Research Board of Canada; The Ministry of Agriculture, Fisheries and Food of the United Kingdom; The Scottish Home Department; The Commonwealth Scientific and Industrial Research Organization, Australia.

4 | Growth Processes in Fish

> "... for understanding production processes there is no adequate substitute for direct information on the underlying relations or mechanisms, such as might be provided by a combination of experimental and field observations" (Paloheimo and Dickie, 1966a).

Chapter 2 mentioned various growth phenomena involving in some instances the whole animal and in others its organs and tissues. This chapter considers these phenomena as they relate to the individual fish.

I. GROWTH CURVES

Brief general consideration to growth curves has been given in Chapter 2; for more detailed biomathematical interpretation there are various other works, such as Graham (1956), Beverton and Holt (1957), Ricker (1954, 1958), Huxley (1932), Medawar (1945) and D'Arcy Thompson (1942). It should, however, be pointed out that, for all organisms, the course of growth may vary in only a few *regular* ways. An organism can grow at an almost constant rate, at a rate that decays or increases with time or at a rate that changes periodically from slow to fast and back again—as D'Arcy Thompson (1942) was among the first to note. Most growth that appears to change irregularly or haphazardly may, when examined at the appropriate "magnification", usually be analysed into one or more growth curves of common types joined end-to-end (see also Chapter 2). Prospects for analysing such curves when they reflect the growth of the whole organism are restricted to studying the effects of modifiers such as space, food, climate or the effects of other organisms. For fuller understanding of the events reflected in changing growth we must consider processes within the growing organism, first noting a few ways of representing growth in fish which have been found useful.

Figure 2.1 showed not only that the sigmoidal growth curve may be expressed as its reciprocal (dy/dt) but that it may also yield a curve of specific growth rate which decays throughout the life of an organism or population. To obtain the data needed for calculation of specific growth rate, measurements of length, weight etc. must be made frequently on individuals or populations. Given suitable data the specific growth rate is derived from use of expressions of the type:

$$YT = t.e^{g(T-t)}$$

where YT is the final size (at time T), and Yt is the initial size (at time t), T and t are given in time units, T being later than t, e is the base of natural logarithms, g is a growth rate known variously as specific, instantaneous, geometrical, multiplicative etc. (Frost and Brown, 1967; see also Brown, M. E., 1946a, 1957; Ricker, 1958). A more convenient form of the expression is:

$$G = \frac{\text{Log e } YT - \text{Log e } Yt}{T-t} \times 100$$

The major advantage of determining specific growth rate is that an investigator is then able to make comparisons of relatively short periods of growth, either between animals (or populations) or at various times in the life of a single animal. The effects of factors such as climate or growth may then become more obvious.

Growth is often treated as if it were a continuous process; as in Fig. 2.1 where smooth curves are used to depict increase in size with time. In that case specific growth rate could be derived from a sigmoidal curve of continuous growth only by differentiation (e.g. Medawar, 1945):

$$\frac{d \text{ Log } W}{dt}$$

But, as already mentioned, growth is commonly a discontinuous process—even degrowth being possible. For fish, it is only "if growth cycles are 'smoothed out', e.g. by using annual data for . . . fish which have annual growth cycles (that) curves showing lengths or weights plotted against age are generally sigmoidal . . ." (Brown, M. E., 1957). It is also because of this that the specific growth rate is a useful determination. As an index of an extended period of changing growth it can furnish values of only limited interest; but if data on changing size can be collected with sufficient frequency, the calculation of specific growth rate can provide a sensitive means for evaluation of a growth pattern (e.g. Frost and Brown, 1967; Ricker and Foerster, 1948; Le Cren, 1951; McFadden, 1961).

II. CONDITION

Another important derivative of growth is what fish population analysts have termed "condition factor", ponderal index or more popularly K (k) factor. In any material body in which, with increasing size, the linear proportions remain constant, weight (if density remains constant) and any bodily linear dimension are related as in the following expression:

$$K = \frac{W}{L^3}$$

As thus determined, K will remain constant no matter how large an animal becomes, providing linear proportions (shape) remain constant. A change of weight at a particular length or change of length, without corresponding change in weight, will alter K. Although this general relationship between length and weight has been valuable in the study of fish populations, its use has been repeatedly criticized. For instance, if one could establish so close an empirical relationship between length and weight for a particular species that K always fell within narrow limits, length could not only be confidently determined from weight (and vice versa)—a great advantage in fishery research—but the weights-for-age of fish populations could be derived merely by determining lengths-for-age by back-calculation from scales etc. In practice, however, even among the members of one population sampled on a single date, there may be considerable variation in condition for any particular fish length and also a change in condition with length (Fig. 4.1). Fish populations often display considerable changes in average condition reflecting normal seasonal fluctuations in their metabolic balance and in the pattern of maturation and subsequent release of reproductive products. Even the state of fullness of the alimentary canal may influence K. In practice, therefore, considerable care is needed in using K to compare the gross nutritional state of different age groups or populations of fish. And in many species significant permanent sex differences in K occur after maturity.

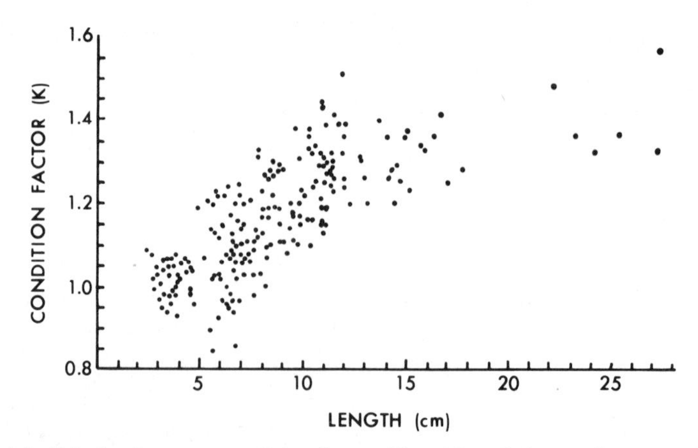

Fig. 4.1. Relation between condition factor, *K,* and length for tench captured in Lake Tiberias, Tasmania, on 12 September 1956. (Redrawn after Weatherley, 1959.)

The procedure of *K* determination has been most effectively utilized in three kinds of population analysis:

(i) In comparing two or more monospecific populations living under apparently similar, or apparently different, conditions of food, density, climate etc.

(ii) In determining the timing and duration of gonad maturation in populations.

(iii) In following the gradual build up or decline of feeding activity over an extended period, or population changes possibly attributable to alterations in the food supply, as reflected in the gross nutritional balance of the fish.

Many workers on trout (e.g. Frost and Brown, 1967) have found it quite convenient to use the simple formula for determination of *K* either as given above or as stated more explicitly for use with the metric system:

$$K = \text{approx. } 1{\cdot}0 = \frac{100 \times \text{weight (g)}}{\text{length (cm)}^3}$$

If the British system of size units is used and it is still required that *K* approximate to unity, the formula (for trout) becomes:

$$K = \text{approx. } 1{\cdot}0 = \frac{1000 \times \text{weight (lb)}}{0{\cdot}427 \times \text{length (in)}^3}$$

There has been some dissatisfaction with this simple formula because it happens to be more suitable for salmonids than for most other groups. For instance the condition factor for eels, thus calculated, is much less than unity, whereas for many fish of more corpulent form the reverse is true. For such reasons Le Cren (1951) sought a formulation of weight/length more generally suited to a wide range of species and body forms. He fully reviewed earlier evidence and this text is largely restricted to his treatment of the problem (but see also Hile, 1936; Allen, 1951; Beverton and Holt, 1957).

Le Cren pointed out that the weight/length relationships in fish may be expressed generally by:

$$W = aL^n \tag{4.1}$$

where W = weight, L = length, a is a constant and n is an exponent "usually lying between 2·5 and 4·0." This expression can also be written:

$$\text{Log } W = \text{Log } A + n \text{ Log } L \tag{4.2}$$

when the weight/length data for fish with the same weight-for-length relationships will fall approximately on a straight line if plotted on log-log paper.

More generally, if a fish maintains one shape as it grows:

$$W = cL^3 \tag{4.3}$$

This is obviously akin to the formula for the K factor already given:

$$\text{i.e. } c = \frac{W}{L^3}$$

When thus calculated, however, the value of c is often an awkward fraction and Le Cren showed that, by finding an average value of c from (4.3) by trial and error, a new K factor could be found for individual populations or species of fish such that:

$$K = \frac{W}{cL^3}$$

where c is a factor intended to bring the value of K close to unity for the majority of individuals. However, Le Cren pointed out that in many cases the cube law fails to apply so that, in any case, $n \neq 3$ for a population as a whole because it will include a range of size classes. He therefore recommended use of the more general $W = aL^n$ to calculate what is termed the "relative condition factor" (Kn) for, as he explained, the weight/length relationship must first be calculated, from formula (4.2), then smoothed mean weights, W, calculated for each age group can be read from an accurate graph. Relative condition factors are then derived by use of the expression:

$$Kn = \frac{W}{\hat{W}}$$

Kn is a measure of the deviation of a given fish from the average weight-for-length for its age group, size group or section of a population or species. K is, by contrast, an individual's deviation from "a hypothetical ideal fish" shape. Le Cren admitted that the choice of which condition factor to compute depends on circumstances. He used Kn in investigating the changes in condition with season and between age groups of Windermere perch. But the use of Kn demands very considerable computational labour and it is doubtful whether the results are usually worth the effort. Certainly, $K = W/L^3$ cannot yield a K very near unity for all species or even for all ages of a particular species, while a relative condition factor near unity can be derived for any group of fish which has a fairly narrow size range by use of $Kn = W/aL^n$, in which case $n \neq 3$ but is derived empirically for each group. The next step is to attempt to assess the different results of using the two expressions for calculating K and Kn respectively.

If K is calculated (weights and lengths in metric units), most individuals of a relatively few species will have a K close to unity even without the use of the constant c. Even for trout, however, K may depart rather widely from unity at certain times of the year, such as spawning time. Other species will have K's greater or less than unity at nearly all times of their lives (e.g. eels, puffer fish). Use of a species-specific constant c of suitable magnitude could, of course, convert these aberrant K's to unity as effectively as in the "chance" case of trout. Indeed, such a constant is applied to obtain a condition factor of 1·0 for trout when using the British system for weights and lengths (see above).

If, on the other hand, Kn is used it is possible to hold its value near unity for all species, populations, age groups etc. but to do so n, as empirically determined, will vary with the different species, populations etc.

It therefore appears that whichever method is chosen, some specification is needed. That this specification is implicit in the power of n where Kn is being expressed does not remove the tedious labour of calculating n separately for each group of fish if Kn is to remain near unity, nor the need to indicate explicitly the size limits for which a given range of c is deemed to apply, if K is to be employed.

If, however, the simple expression:

$$K = \frac{W}{L^3}$$

is employed do we not have a satisfactory index of condition? Suppose the K of a certain species does depart more or less greatly from unity. Knowing the magnitude and direction of the departure we can set up standards which vary, relatively, no more or less than the variation around unity for trout. There seems no intrinsic virtue in a value of unity for K and if we only manage to replace the simple computation of K with one demanding laborious readjustments using varying values of c or n, then why bother? The reader is also referred to Beverton and Holt (1957) and Nicholls (1957).

Most trends in condition with growth of a given species are gradual, whether shown as changes in K in the "traditional" calculation or in the values of c or n if K is to be kept near unity. Thus, though these variables may increase or decrease with size they are rarely so great that K belongs to different orders of magnitude in successive size groups. There are cases, however, where fish undergo considerable metamorphosis during larval development which would be reflected in a rapidly changing K, indicating a high order of difference between pre- and post-larval fish. The reason this phenomenon has not often been recorded quantitatively is because larval life is normally of short duration and larval fish are frequently too small to measure and weigh accurately in the numbers needed for a population survey. In certain species marked morphological change accompanies growth after the larval stages—e.g. in sailfish and dealfish (Greenwood, 1963), but even to monitor changes in condition during these periods would call for many measurements of weight and length at relatively short intervals.

Suppositions that alterations in condition during growth are produced by changes in bodily proportions rather than alteration of fat depots, gonad size etc. should be tested. Otherwise the ecological significance of condition changes could be seriously misconstrued. Weatherley (1959) found that depth and breadth of tench *Tinca tinca* bore a linear relationship to body length over a fairly wide size range (2-27 cm), yet K ($=W/L^3$) showed a considerable tendency to increase with total length (Fig. 4.1). So it appears that mean density of body increased as the fish became larger.

Brett *et al.* (1969) showed that proportions of water, protein and fat in the bodies of fingerling sockeye salmon *Oncorhynchus nerka* can be greatly affected by size and composition of ration and by water temperature. Thus considerable differences in density may be expected within a species.

Let us now set aside differences between methods of determining condition and briefly consider typical trends in condition in various fish populations.

Hile (1936) reviewed others' results and added further data on condition in cisco *Leucichthys artedi* living in North American lakes in which clear differences were detected between populations. Allen (1951) examined trends in condition in various year-classes of the trout population living in several sections of the Horokiwi River and found few consistent differences, although there was a tendency for condition to be rather higher in summer which coincided with more rapid growth at that time. Allen also showed that floods reduced growth with an accompanying fall in condition. Allen (1938) had previously suggested a similar relation between condition and growth in the trout of Windermere and in the growth of very young salmon *Salmo salar* in the Eden and Thurso River systems.

Le Cren (1951) investigated "relative condition" throughout the year in the Windermere perch population in which the main differences between the patterns of change in mature males and females and immature fish were attributable to the sexual cycle–i.e. change in gonad size (Fig. 4.2). There was a massive and rapid loss in relative condition in mature females at spring spawning when large quantities of ova were released. The maturation of gonads in perch has the effect of transforming a monophasic annual cycle of condition (seen in the immature fish) into a diphasic one, with spring and summer peaks of condition. In mature tench (20-30 cm long) in Tasmania, Weatherley (1959) found a simpler condition cycle. A loss of condition follows release of ova and sperm, but because this occurs in summer rather than spring, as in the perch, there is apparently no time for the tench to regain condition again in a particular year so, after spawning, condition declines to its winter low (Fig. 4.2).

III. TEMPERATURE AND GROWTH

There have been relatively few critical investigations of growth in which fish fed to satiety or to various specified levels have been maintained at different constant temperatures.

Young guppies (*Lebistes*) were held in 1 gallon jars in ¼ strength

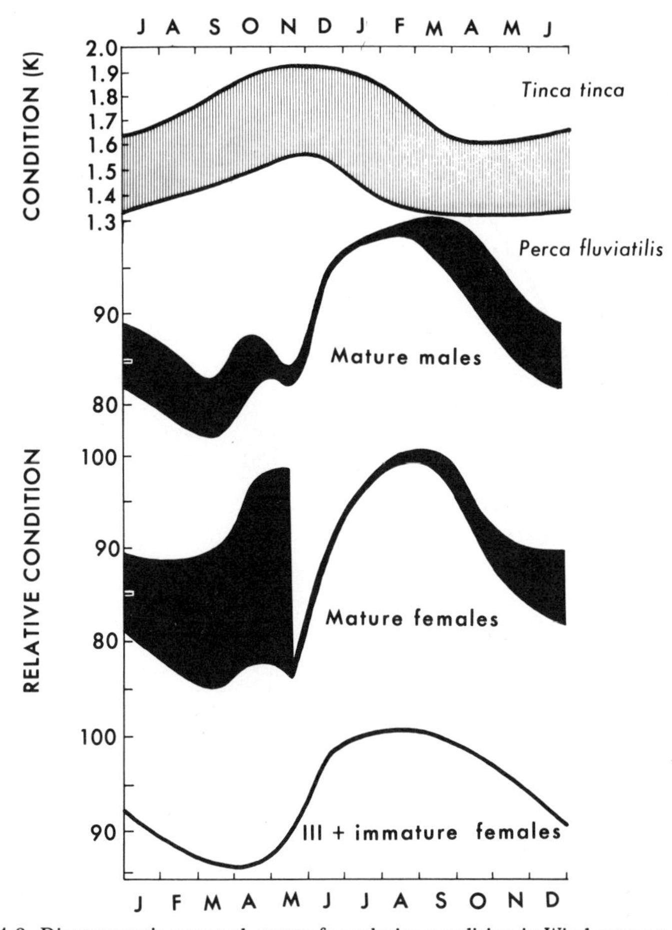

Fig. 4.2. Diagrammatic seasonal curves for relative condition in Windermere perch, with and without gonads. The solid black area represents the gonad weights; the upper edge of the black the condition with gonads, the lower edge condition minus gonads. (Redrawn after Le Cren, 1951.) At top of figure condition cycle of tench in Lake Tiberias is shown for comparison. Shaded area between lines covers the range of condition found. (Modified after Weatherley, 1959.)

sea water and in "freshwater" at temperatures of 20°, 23°, 25°, 30° and 32°C (Fig. 4.3); 23° and 25°C were most favourable for growth, which was less rapid above and below these temperatures. The fewer experiments done in ½ strength sea water confirmed this trend. Lack of information on feeding, and variation in population density between jars of from 4 to more than 16 individuals, makes it difficult to interpret these findings.

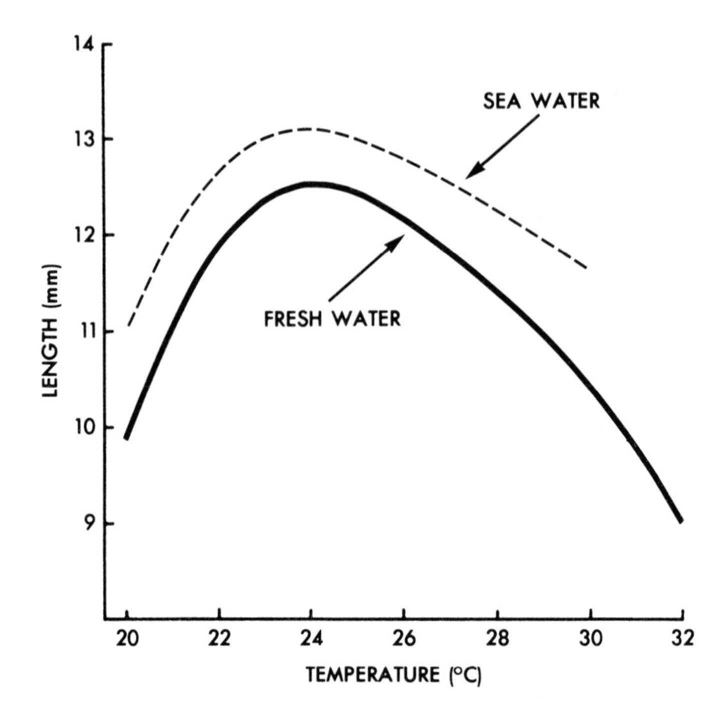

Fig. 4.3. Curves of average lengths of 40-day-old guppies in 1/4 strength sea water at various temperatures. (Modified after Gibson and Hirst, 1955.)

Brown, M. E. (1946a, b, c) performed pioneer work on trout and these studies have been subsequently reviewed and amplified (Brown, M. E., 1957; Frost and Brown, 1967). Brown's most significant finding was that 2 year old brown trout kept at temperatures of 4·5°, 8°, 12°, 15°, 18° and 20° C, had specific growth rate maxima at 7°–9° C and at 16°–19° C. Maintenance requirements increased along a sigmoidal curve with increasing holding temperature (Fig. 4.4). This curve was interpreted as meaning that while standard† metabolism increased exponentially up to 20° C and beyond, activity increased to a maximum between 8° and 18° C, then declined. Figure 4.5 depicts Brown's tentative explanation of how these several variables could interact to produce two peaks of specific growth rate. Other experiments led Brown, M. E. (1946b) to suggest an essentially endogenous annual growth cycle in trout. Frost and Brown (1967) elaborated these ideas by reference to natural events in the life of trout in British waters, suggesting the existence

† Standard metabolism is explained on p. 104

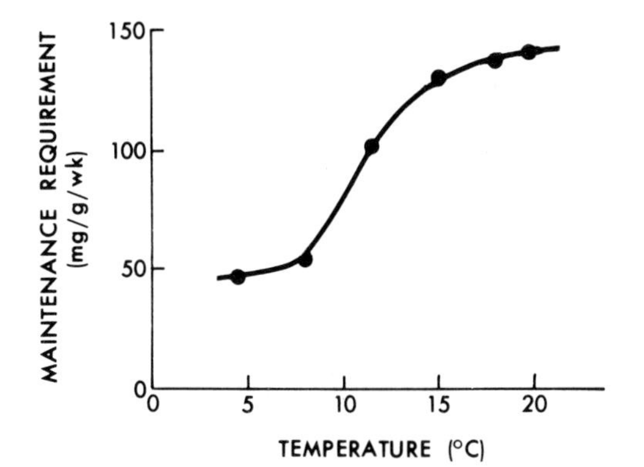

Fig. 4.4. Calculated maintenance requirements for brown trout of mean weight 50 g held at various temperatures. (Redrawn after Brown, M. E., 1957.)

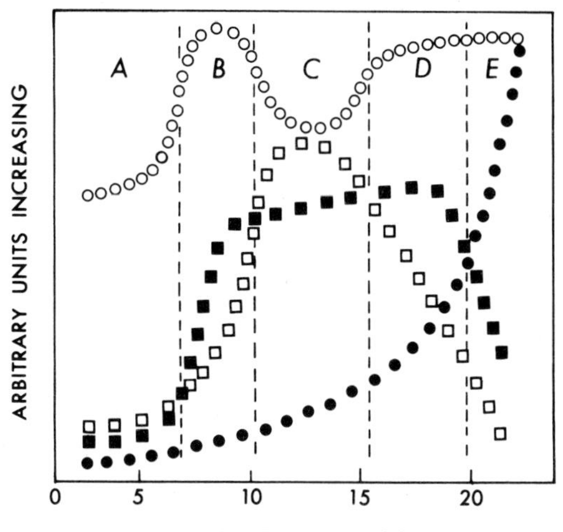

Fig. 4.5. Scheme indicating how growth rate maxima occurring at two different temperatures could result from a differential effect the temperature had on the amount of food eaten (■), the activity (□), and the standard metabolic rate (●), of trout. Vertical lines divide the temperature range into five regions within three of which (A, C and E) specific growth rate would be low; in regions B and D it would be high. The change in efficiency of food utilization (○) with temperature would exaggerate the growth rate maxima and minima. (Redrawn after Brown, M. E., 1957.)

of an endogenous growth rhythm synchronized so as to take metabolic advantage of the times of the year most favourable for food and temperature conditions.

Swift obtained results at first seeming to confirm Brown's findings. He held trout, fed to satiety, in stewponds in which temperature, chemical composition of water and day-length showed natural fluctuations (Swift, 1955). Specific growth rate showed two annual maxima between 8° and 12°C and between 15° and 16°C which, despite different experimental conditions, agreed with Brown's results. Swift (1961) described growth of trout in a stewpond, a tarn, a netted-off section of Windermere and also held in constant temperature tanks. Swift critically evaluated his own earlier observations and recast his views in approximately the following form. When held at several constant temperatures, 8°, 12° and 16°C, trout grow most rapidly at 12°C. Specific growth rate increases with temperature during the spring up to about 10°–12°C when growth becomes inversely proportional to temperature, so that during summer growth rate falls. It rises again in the autumn as water temperature falls but at a higher temperature than in the spring and at about 14°C, growth rate becomes proportional to water temperature.

Later Swift investigated the crucial question of endogenous growth rhythms, which Brown, M. E. (1946b) considered she had demonstrated in 2 year old trout held under constant environmental conditions at 11·5°C. Swift (1961) had already queried this in pointing out that the high variability of growth rate among the relatively small number of fish Brown used could possibly account for the apparent growth pattern. Swift (1962) held pairs of trout at 5°, 8°, 10° and 12°C for 10 months (December to September) and found that individual growth rates varied according to the time of the year and that the variations were not in phase. Later still Swift (1964) concluded that char (*Salvelinus alpinus willoughbii*) under constant temperature conditions had, like trout, a single maximum of specific growth rate occurring at 12°–16°C. At 18°C growth rate was as low as at 4°C. Frost and Brown (1967) recently admitted the controversial nature of the relationships between growth and temperature, while claiming that good growth can be expected to occur only in the temperature range 7°–19°C.

Let us now turn to temperature and fish growth in wild populations. Weatherley and Lake (1967) compared trout growth in waters of the south-eastern Australian mainland with that typical in Tasmania, New Zealand and the British Isles. Figure 2.5 indicates the rapid growth of trout in New South Wales rivers. In Tasmania they grow more slowly and in Britain usually still more slowly. Although not given here, New Zealand trout have growth rates comparable to those in Tasmania. Basing their interpretation partly on Swift's findings (1961) and partly on general climatic differences, Weatherley and Lake suggested that more favourable thermal conditions were largely responsible for the good growth in Australian waters. The evidence for this was admittedly indirect, based on overall climatic differences rather than precise comparisons of temperature conditions in different waters:

> . . . the general comparisons of growth are for three very large regions, so that differences due to special local factors are rendered relatively unimportant, while general climatic conditions will be the most consistent differences distinguishing them. Accordingly, we tentatively conclude that general regional differences between trout growth rates can be accounted for mainly in terms of climatic differences. It is perhaps of little practical importance whether the effect is directly produced by temperature or indirectly through the effects of temperature on food production.

Here, Nicholls' (1957) experiment should be mentioned. In Tasmania, he kept three trout, fed to satiety once each day, for three years in hatchery troughs. The hatchery water came from a neighbouring river, so that diurnal and seasonal temperature changes were essentially natural. Against Swift's (1961) experimental findings and Frost and Brown's (1967) supposition that trout growth falls to near zero in midsummer, Nicholls' trout grew throughout spring, summer and autumn, even when water temperature reached 20° C. Growth ceased in winter at temperatures below 4·5° C. Allen (1951) also reported growth in trout in the Horokiwi during summer when water temperatures reach nearly 20° C in midsummer.

Swift (1961) indicated that past thermal experience might condition the critical temperatures for growth in fish. On the basis of this idea and Nicholls' findings, Weatherley and Lake (1967) suggested that if trout in Australia were to some degree thermally adapted to the climate then such an effect could reasonably be expected to be emphasized in New South Wales where temperature conditions are warmer in trout streams than in any of the other environments mentioned above.

Le Cren (1958) found that Windermere perch exhibit most of their annual growth from June to September—the time when water in the littoral epilimnion (the main abode of perch in summer) exceeds 14° C. Temperature records from 1935, converted to degree-days in excess of 14° C, showed a strong correlation with year-to-year fluctuation in growth of various year-classes. Figure 4.6 shows the least squares regression line that accounts for this relation among adult fish (Le Cren, 1958). Le Cren ascribed two-thirds of the year-to-year variations in growth to temperature, notwithstanding the important effects of changing population density and other factors, described later.

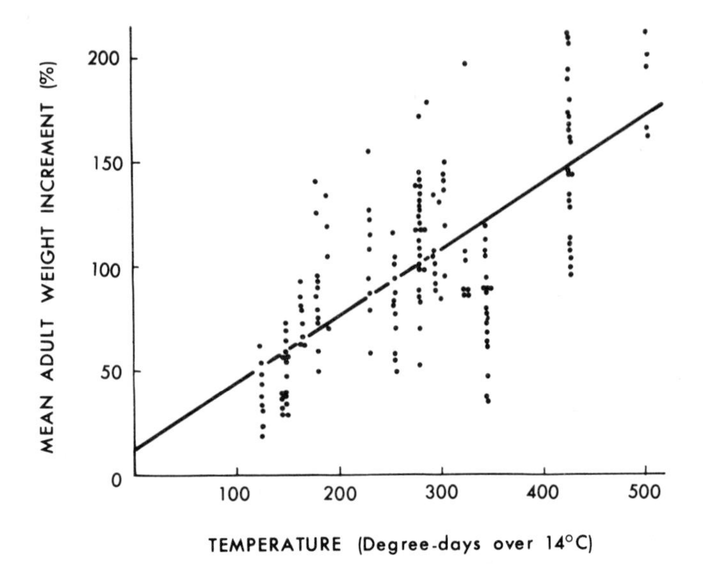

Fig. 4.6. Percentage of average annual weight increment in adult perch in Windermere when adult increment for each year-class for each year is plotted against summer surface water temperature. Points are for individual year-classes of each sex. Line is least-squares regression. (Simplified after Le Cren, 1958.)

Allen (1940, 1941) sampled stages (i.e. in fresh water) of salmon *Salmo salar* in the river Eden in England and the Thurso River System in Scotland. Frequent sampling enabled Allen to depict clearly the actual course of growth in wild populations and he concluded that temperature did not so much control growth as trigger it, enabling it to accelerate in the presence of adequate food, when temperature had reached a critical value. He assessed food abundance as the major controlling influence, even though plentiful food did not in itself lead to rapid growth; for that, water temperatures had to exceed 7° C. As confirmation of this view of growth factors, when temperature fell below 7° C at the end of summer, growth in young salmon declined rapidly even in the presence of relatively plentiful food.

Le Cren's (1958) ideas on growth of perch in relation to temperature in Windermere are in interesting contrast to the above. He believed that although the entire biota, including those organisms used as food by perch, was undoubtedly affected by temperature, since perch of all sizes were affected in the same rapid way, the effect of temperature on growth was probably direct–i.e. essentially physiological.

Knowledge of the upper and lower limits of temperature and of optimal temperatures for the growth of most species of fish is, fairly obviously, very incomplete. Innumerable species of tropical fish grow actively at temperatures much above those optimal for many temperate species: the world's highly successful tropical fisheries, including those based on pond culture, frequently depend on this fact (e.g. Schuster, 1952; Hickling, 1961, 1962). Many tropical fish grow at temperatures up to and exceeding 30° C. At the other end of the scale fish can grow rapidly in polar waters, and Wohlschag (1961) has even described growth in a benthic nototheniid fish *Trematomus* at temperatures under the ice, in McMurdo Sound, of about –2° C. This range shows that generalizations about the effects of temperature on fish growth that stray beyond the confines of particular groups are apt to be meaningless. The true position appears to be that in the thermal governance of growth, as in other matters, fish display an impressive variety of adaptations. It seems clear that some fish employ the temperature changes in the natural environment as indicators of the impending arrival of the most suitable seasons for growth. Thus, in temperate regions growth may proceed rapidly in spring when food is likely to be plentiful. Nikolskii (1969) gives useful additional data and discussion of this problem.

In Chapter 2, Fig. 2.8 showed the fate of net energy from the ingested food indicating that it is shared by the demands of maintenance, new material and the "work of growth". This last value is the metabolism which accompanies accretion of new living material and which naturally changes in rate as growth rate itself changes. Since, in fish, growth rate is shown to vary with temperature we can expect that "work of growth" will be temperature-dependent. However, the order of magnitude of this variable has yet to be assessed.

IV. METABOLISM

A. Food and Growth

Figure 4.5 shows Brown's essentially metabolic scheme to account for the results of her experiments on trout growth. Although it has long been realized that fish growth curves can be regarded as resultants of metabolism—anabolic excess over catabolic loss—full application of this approach to growth analysis has been slow. It seems possible that the expanding use of mathematical methods for describing and fomulating growth led to this neglect.

Bertalanffy pioneered in showing fish biologists the possibility of developing metabolic models of growth, although Pütter may have first stated the essential proposition. Bertalanffy's concepts underlie much of what follows even though not often explicitly stated. Students of growth should refer particularly to Bertalanffy's (1960) comprehensive recent monograph on this topic.

Warren and Davis (1967) have discussed the utilization and losses of energy by an animal as shown in the formulas and notations of Ivlev and Winberg.

Ivlev's formula (Ivlev, 1939a, b, 1945) is:

$$Q = Q^1 + Q_r + Q_t + Q_v + Q_w$$

where energy values are:

Q, for food consumed
Q^1, for matter incorporated into body (growth)
Q_r, for faeces and nitrogenous excretion
Q_t, for primary heat
Q_v, for external work (activity)
Q_w, for "internal" work.

Winberg's (1956) balanced energy equation is:

$$Qc = 1{\cdot}25\ (Qr + Qg)$$

where

Qc = energy of ration
Qr = energy of metabolism
Qg = energy of weight increase.

Winberg (1956) pointed out that it is difficult to understand what Ivlev means by "primary heat". Warren and Davis (1967) have criticized Winberg's formula principally because his idea of routine metabolism and the relation it bears to activity and nutritional levels of fish requires clearer definition. Their own formula is based on the notation (Fig. 4.7) which bears an obvious kinship with Fig. 2.8 (from Needham, 1964; but see also Kleiber, 1961 and Davies, 1964).

From the scheme in Fig. 4.7 we can write:

$$Qc - Qw = Qg + Qr \qquad (4.4)$$

moreover

$$Qr = Qs + Qd + Qa \quad \text{(also from Fig. 4.7)}$$

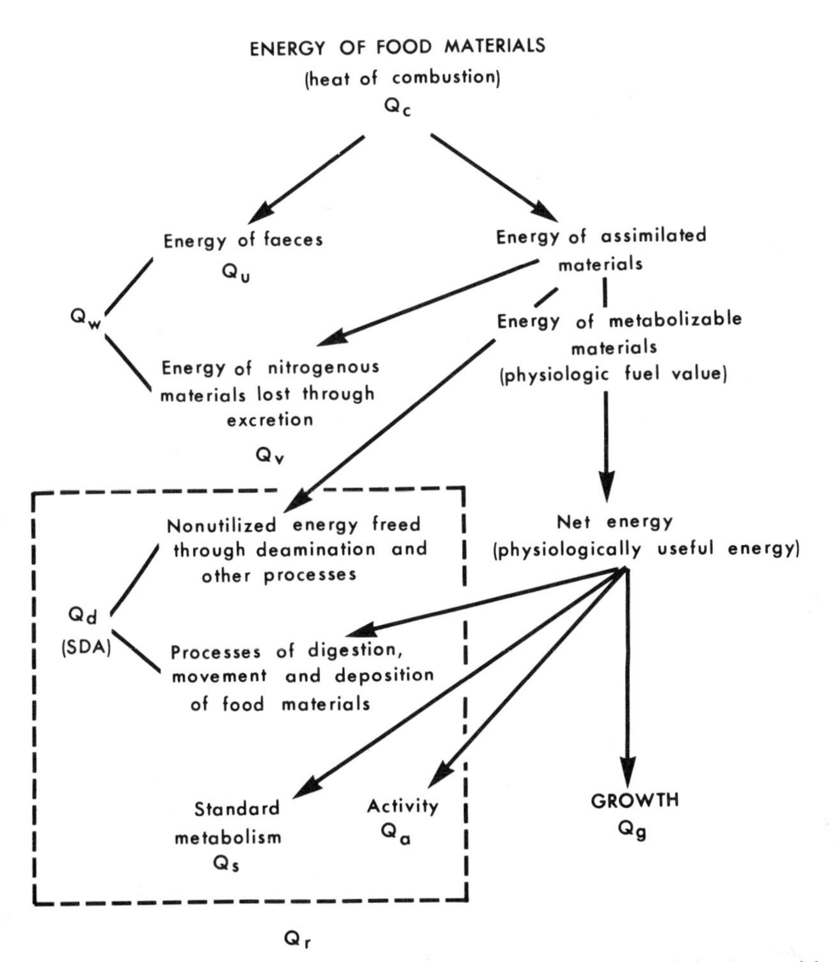

Fig. 4.7. Categories of losses and uses of the energy of consumed food materials. (Redrawn after Warren and Davis, 1967.)

so that

$$Qc - Qw = Qg + Qs + Qd + Qa \quad (4.5)$$

By using energy equivalents for all the variables in the above equations Warren and Davis avoid, in a sense, the problem discussed by Pandian (1967) who observed that calculations of the metabolism of foodstuffs based on wet tissue weights are of dubious value, because of the varying proportion of fats, carbohydrates and proteins in different foods and their varied metabolic fates.

Warren and Davis (1967) urged the increasing use of bomb calorimetry (see also Davies, 1963, 1964, 1966, 1967) in fish growth studies, reinforcing the objection of Pandian (1967). Dry weight measurements, not very informative in themselves, can have their value enhanced considerably if the protein and fat contents of the weighed organisms—from which energy equivalents can be computed from normal caloric conversion factors—are known. Knowledge of the efficiency of digestion is also desirable (Pandian, 1967: Molnar, 1967; Windell, 1967).

Warren and Davis converted equation (4.5) to rate terms by assuming all energy losses proportional to the same power function of weight (0·8); thus:

$$Ac - Aw = Ag + As + Ad + Aa \tag{4.6}$$

when

$$Ai = \frac{Qi}{W^x t}$$

where

W, is mean weight or energy value of body
t, is time in days
x, is some mean power of W
$i = c, w, g, s, d$ or a.

The structuring of the energy budget in terms of food consumption and loss and use of energy is illustrated in Fig. 4.8 based on data of H. Sethi, for the cichlid *Cichlasoma bimaculatum* (Warren and Davis, 1967). Particularly noteworthy are the very high values of the specific calorigenic action of food throughout the entire temperature range employed. Above 32° C food consumption rose significantly, the specific calorigenic action coming to represent the largest component of the energy values of the food consumed. The increase in specific calorigenic action and in faecal wastage were not associated with an increase in scope for growth. It seems more than likely that the increased activity of fish at high temperatures would tend to reduce their scope for growth, since even when fed freely such fish often tend to lose weight and show widespread degenerative changes (Cocking, 1957, 1959a,b; Weatherley, 1963). Indeed, the specific calorigenic action, which Sethi evidently computed by adding growth, metabolism of starved fish (standard metabolism) and faecal wastes and subtracting the sum from food consumed, may be too high—especially at high temperatures where the cost of activity might not have been allowed for.

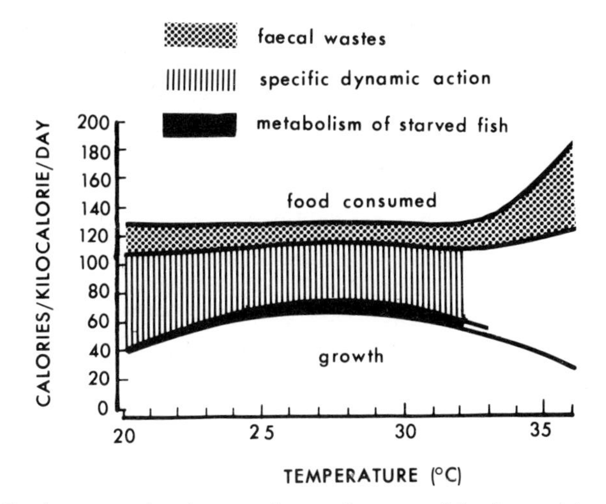

Fig. 4.8. Food consumption, losses and uses of energy of food materials and scope for growth of young *Cichlasoma bimaculatum* in calories per mean kilocalorie of fish biomass per day, in relation to temperature. Fish fed *ad libitum* for 14 days. Categories of faecal wastes and specific dynamic action include small percentages of nitrogenous wastes. Estimates of metabolism of starved fish were based on experiments with fish larger than those used for other estimates and have been adjusted for size differences. From data of H. Sethi. (Redrawn after Warren and Davis, 1967.)

Warren and Davis (1967) employ the formulas for gross efficiency and net efficiency† (as defined in Chapter 2), using the terms gross and total efficiency interchangeably. The formula for total efficiency is:

$$\mathrm{E}t = \frac{G}{I} \tag{4.7}$$

where G = growth, I = food intake.

For net efficiency, also called partial growth efficiency

$$\mathrm{E}pg = \frac{G}{I-M} \tag{4.8}$$

where M = maintenance ration.

† Formulae such as these can also be expressed directly in terms of energy exchanges or outputs; e.g.:

$$\text{Total efficiency} = \frac{A}{H}$$

where

A = heat equivalent of work performed
H = metabolic rate x duration of work

See Kleiber (1961) for other examples.

Warren and Davis also gave a formula for "partial maintenance efficiency":

$$Epm = \frac{Lp}{Ip}$$

where Lp = tissue loss prevented
Ip = part of ration preventing loss.

In aquarium experiments with yearling sculpins *Cottus perplexus*, Warren and Davis compared autumn growth with winter growth. Net efficiency against food consumption (Fig. 4.9) showed a slight

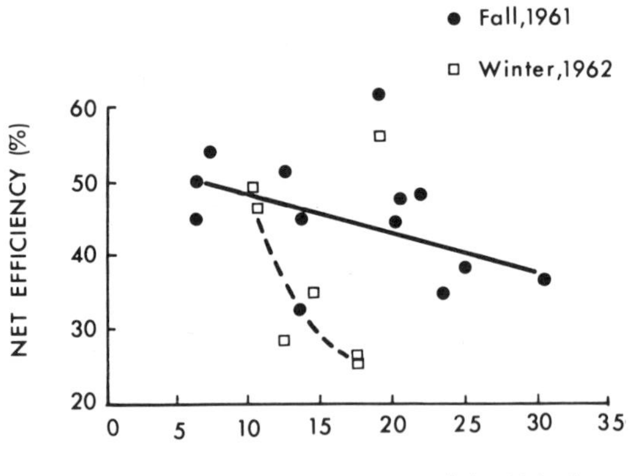

Fig. 4.9. Relationships between rate of food consumption and net efficiency of food utilization for growth of yearling sculpins (*Cottus perplexus*) in aquarium experiments during autumn 1961 (•) and winter 1962 (□). (Redrawn after Warren and Davis, 1967).

decline with increasing ration size, maximum gross efficiencies being achieved at two-thirds maximum ration. In winter, sculpins had net efficiencies similar to those in autumn on rations a little higher than maintenance level, but as rations rose net efficiency declined sharply. As Warren and Davis indicated, it is important that on low winter rations temperate fish should possess the means to utilize the food they ingest as efficiently as at warmer periods. After all, fish able to maintain weight or anyway to lose weight slowly, when feeding activity is very limited, would clearly seem to have advantages over those lacking such ability. Ivlev (1961) found in experiments that when fish are starving there exist considerable interspecific differences in ability to conserve weight.

Warren and Davis (1967) (citing data of R. Brocken) showed that cutthroat trout *S. clarki* had a higher growth efficiency in late winter than in spring—at least when receiving low rations.

Paloheimo and Dickie (1965, 1966 a, b) recently discussed the manner in which food energy contributes to growth in fish. Their interesting presentation unfortunately says little about the fish population living in the wild. The analysis was further developed by Beamish and Dickie (1967).

The original Bertalanffy growth equation (1938, 1957), taken by Paloheimo and Dickie (1965) as their starting point, is:

$$\frac{\mathrm{dw}}{\mathrm{d}t} = Hw^m - D\mathrm{w}^n \tag{4.9}$$

where H and D are coefficients of rate of anabolism and catabolism and W = body weight ($m = 2/3$, $n = 1$).

At the outset, Paloheimo and Dickie (1965) prefer as a growth formula the basic energy equation of Winberg (1956):

$$\frac{\Delta W}{\Delta t} = R - T \tag{4.10}$$

where ΔW is the energy equivalent of growth during a period of time t and is the energy acquired in the rations R, minus the total energy expended in metabolism T, measured during the period of time (this equation is given above in a slightly different form). Paloheimo and Dickie (1965) state that, "In place of the abstract concepts of anabolism and catabolism, which necessitate a somewhat arbitrary specification of their relation to body weight, we thus have three quantities which are clearly open to experimental study."

For fish:

$$T = \alpha W^{\gamma}$$

i.e.

$$\text{Log}T = \log\alpha + \gamma\log W \tag{4.11}$$

(also known as the T-line)

where

W = body weight

α = constant for the metabolic rate

γ = metabolic rate of change with change of body weight (i.e. the "slope of the log line").

From their analysis of various experiments Paloheimo and Dickie (1965) claim a value of γ approximating to 0·8, so that from formulas (4·10) and (4·11)

$$\frac{\Delta W}{\Delta t} = R - \alpha W^{\gamma} \tag{4.12}$$

This permits derivation of a growth equation if data are

forthcoming on changes in R. It is also clear from (4·12) that $R > \alpha W^{\gamma}$ is a condition if growth is to occur.

Growth in the present immediate context should be taken in Bertalanffy's (1960) sense to mean increase in body bulk, "production of materials ... specific for the living system ... produced by multiplication of already existing protoplasm."

Paloheimo and Dickie (1965) state that as fish grow they may get all their required food:

(i) without increasing either speed of swimming or distance covered

(ii) by increasing their time of swimming (speed constant)

(iii) by increasing their speed (time constant).

In practice (ii) and (iii) may frequently be combined by force of circumstances; (i) is likely only in the presence of abundant food. In (ii) and (iii) the energy cost of the work needed to acquire food (rations) at a rate proportional to W^{γ} increases faster than W^{γ}, so a situation can readily be envisaged in which the activity of the search for food could itself curtail further growth at any size at which this activity became excessive. In fact, however, many fish which do not obtain their food from the base of the production pyramid (plants), or which do not live in the presence of abundant food animals, change their diet as they grow (Chapters 5 and 7). This has long been supposed to indicate the need of fish to conserve that component of their metabolism which goes into activity. Other methods of improving the efficiency of feeding–behavioural or through morphological specializations–are considered in Chapters 5 and 7. The above equations could perhaps be held to make quantitative physiological sense of such natural history. However, Paloheimo and Dickie (1965) indicate that such considerations do not at present "provide an unequivocal basis for a growth equation", which must rather be reached through actual data on the changing metabolic needs dictated by increase in size.

According to Paloheimo and Dickie (1965):

$$\frac{\Delta W}{\Delta t} = K1 \quad \text{or} \quad \frac{\Delta W}{a'R\Delta t} = K2$$

depending on whether we are considering total rations $R\Delta t$ or "assimilated" rations $a'R\Delta t$; that is, K is the efficiency of the use of food by the fish or "growth coefficient" of the first or second order (Ivlev, 1945) or, to use our earlier terminology, may represent total efficiency or net efficiency. Paloheimo and Dickie (1965) expressed this relationship in the form of three interchangeable equations:

$$\frac{\text{Log } \Delta W}{R \Delta t} = -a - bR$$

$$K1 = \frac{\Delta W}{R \Delta t} = e^{-a-bR} \tag{4.13}$$

$$\frac{\Delta W}{\Delta t} = Re^{-a-bR}$$

where a and b are constants; the other symbols are as given earlier. Paloheimo and Dickie consider growth efficiency to be maximal at low feeding levels when:

$$\frac{\Delta W}{R \Delta t} = e^{-a}$$

The growth efficiency decreases by a constant fraction given by e^{-b} for "each unit increase in amount consumed per unit time, and is independent of the size of the fish."

Growth rate is maximal when $bR = 1$ and may be expressed as:

$$\frac{\Delta W\text{max}}{\Delta t} = \frac{1}{be^{a} + 1}$$

Equations (4.13), describing growth efficiency, apply only for a given food type and range of sizes of ration and fish body. But growth efficiency is apparently independent of body weight, the same "absolute" ration leading to the same "absolute" growth rate irrespective of fish size (Chapter 2 discusses relative growth efficiency of rabbits and cattle). When food types, ration sizes and body sizes exceed a certain range, it may become necessary to organize data so that a series of K lines can be derived if the results are to be made more meaningful.

The next question considered by Paloheimo and Dickie (1965) concerns interactions of food metabolism and growth. From (4.11) and (4.12) above:

$$T = \alpha W^{\gamma} = R - \frac{\Delta W}{\Delta t}$$

then from (4.13) we have

$$T = R(1 - e^{-a-bR})$$

Since $T = \alpha W^{\gamma}$ this means that we may calculate rations for a fish of size W to remain at a given level of total metabolic expenditure from

$$\alpha W^{\gamma} = R(1 - e^{-a-bR}) \tag{4.14}$$

Paloheimo and Dickie (1965) found that b and R may have a wide range of values and, as these actually determine growth and limit the size, they have described a procedure for standardization of growth curves.

At maximum growth rate $bR = 1$ so, on a revised scale of ration measurement, rations may be taken as units of $1/b$.

Then $bR = r$ (r = rations on the new scale).

Since $R = 1/b$, body weight at $r = 1$, the corresponding scale for measuring weight should be

$$\left(\frac{1-e^{-a-bR}}{\alpha b}\right)^{\gamma}. \qquad (4.15)$$

For simplicity (e^{-a} is explicit in the equation)

$$\omega = (\alpha b)^{1/\gamma} W \qquad (\omega = \text{rescaled weight})$$

from (4.14)

$$\omega^{\gamma} = \gamma(1-e^{-a-bR}). \qquad (4.16)$$

The data for these considerations of Paloheimo and Dickie (1965) were given in their two subsequent papers (Paloheimo and Dickie, 1966a, b), which analyse critically a number of studies of fish growth and form an inseparable background to their formal discourse on growth principles (see also Beamish and Dickie, 1967).

Winberg (1956) proposed an equation for growth of individual fish:

$$pR = T + \frac{\Delta W}{\Delta t}$$

where

R = rations consumed
p = a correction factor to convert R to rations assimilated
T = total metabolism
ΔW = weight change
Δt = the period of time for weight change.

The variables R, ΔW and T are expressed in common energy units (e.g. calories). T can, in theory, be derived from oxygen consumption though in practice there may be difficulties. Whether such a gross formulation, in which energy input equals energy dissipation plus energy storage, can be of far-reaching utility will be considered later. Certainly, to be able to carry out actual computations in such a form, which at least deals with fairly readily measurable parameters, would be a considerable advantage.

In their critical review of the scattered experimental data Paloheimo and Dickie (1966a) discussed various values for the "weight exponent" γ in the equation

$$T = \alpha W^{\gamma}$$

or

$$\text{Log } T = \text{Log } \alpha + \gamma \text{ Log } W$$

They concluded that, allowing for the varied techniques and species used and the varied levels of activity, about 0·8 is a reasonable value for the weight exponent γ of this relationship among a majority of fish under more or less constant, mainly non-stress, conditions (see also Warren and Davis, 1967). They also concluded that the constant α (in equation 4.11), which expressed the level of metabolism, is more labile, apparently affected by temperature and specific calorigenic action of food. Furthermore, it is not easy to obtain the consistently quiet conditions required for the measurement of standard metabolism.

In considering the physiologically useful ration–that part not voided in faeces or excretion–it seems that for most purposes p could vary considerably in value without much affecting γ. Thus the equation

$$T = pR - \frac{\Delta W}{\Delta t} \tag{4.17}$$

if

$$K = \frac{\Delta W}{pR\Delta t}$$

(i.e. efficiency of food utilization) remains constant as rations, R, increases, may be written

$$T = \alpha W^{\gamma} = \frac{\Delta W}{\Delta t}\left(\frac{b}{k} - 1\right)$$

As p changes it will affect T rather than γ. The utilization coefficient, p, is about 0·8 as empirically determined.

Paloheimo and Dickie (1966a, b) first analysed Pentelow's (1939) experiments on brown trout kept in tanks, some fed *ad libitum*, others on a maintenance ration; i.e. the food supply was continually and carefully adjusted so as to hold fish weight nearly constant. Pentelow's trout were fed live *Gammarus*; Paloheimo and Dickie (1966a) calculated the energy equivalent of the amount consumed. Figure 4.10 (top) shows the resulting regression lines. Dawes' (1930-31a, b) data for 2 year old plaice (*Pleuronectes platessa*) were

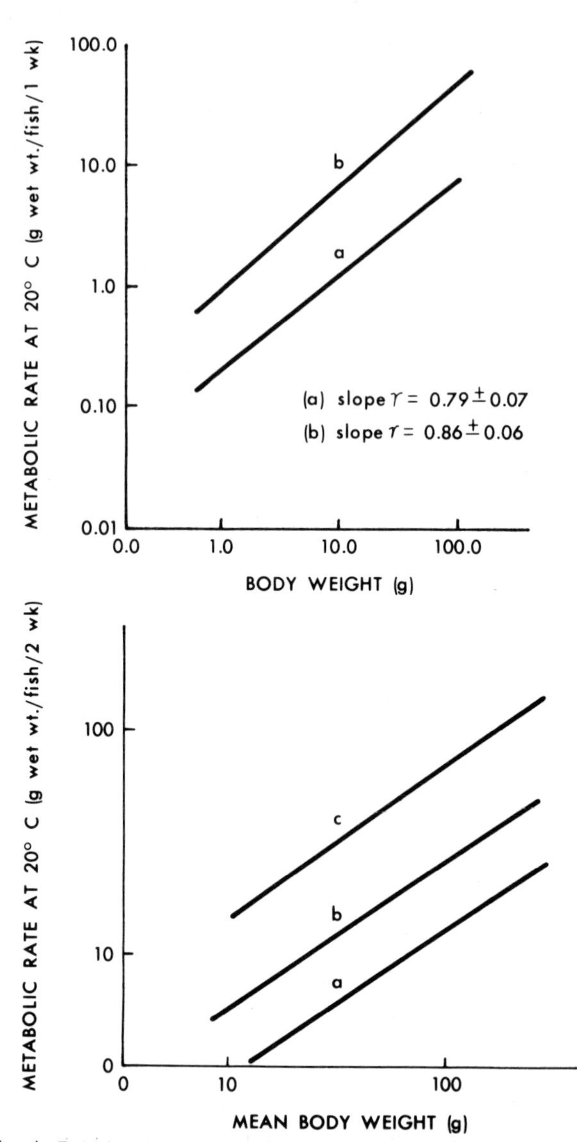

Fig. 4.10 (top). Relation between total metabolic rate and body weight in individual brown trout held at different feeding rates (data of Pentelow, 1939). Line (a)—maintenance rations; line (b)—*ad libitum.* Metabolic rate was obtained by use of Winberg's energy equation ($T = 0{\cdot}8R - \Delta W/\Delta t$) assuming fish and rations to have the same calorie content per unit weight. Temperature adjustment was made from a multiple regression equation given by Paloheimo and Dickie. (Simplified; individual values omitted.)

Fig. 4.10 (bottom). Relation between metabolic rate and body weight of North Sea plaice at 20° C at three different feeding levels. Line (a)—maintenance rations; line (b)—"intermediate" rations; line (c)—*ab libitum* (data of Dawes, 1930-31a, b). The points on which these lines are based are omitted for simplicity but were means of individuals or of groups reared together and of both sexes. (a) and (c) are partial regression lines. Corrected for temperature as in Fig. 4.10 (top).

Both these figures are simplified after Paloheimo and Dickie (1966a.)

considered next. Dawes kept individual plaice in cages, fed on pieces of fresh *Mytilus* at three levels of abundance: *ad libitum,* maintenance and intermediate. Figure 4.10 (bottom) shows that the results resemble those for trout though they differ in magnitude. Moreover, γ values were different from those for trout, perhaps somewhat anomalous. Kinne's (1960) experiments on desert pupfish *Cyprioden macularius* were also examined, with attention directed to the effects of temperature on rations (metabolic expenditure). The fish were fed at a rate less than *ad libitum,* perhaps explaining the sharp decline in metabolism at 35° C for the larger specimens. The food supplies were apparently limiting growth at 35° C; the fall in metabolic rate may have indicated that demand for food was exceeding supply. Otherwise, the γ slope was similar to other results.

Paloheimo and Dickie (1966b) studied a number of other examples from the scattered literature on fish feeding, making due allowance where possible for differing experimental procedures and the more dubious results. Two essential points emerged from these examples. First, at higher temperatures the level of metabolism (α) of fish of various sizes seems to increase, at least in short-term studies. Second, serious departures of γ from a value of about 0·8 seem largely explicable in terms of differences in experimental procedures, sampling error etc. It seems, in addition, that unless fish are fed *ad libitum* (rather than, for instance, to "satiety" at one meal a day) the measured effects of temperature on growth (e.g. Gibson and Hirst, 1955; Brown, M. E., 1946a, b, c; Swift, 1961, 1962, 1964; Nicholls, 1957) may be of doubtful significance.

Paloheimo and Dickie (1966b) investigated effects of ration size on growth efficiency, considering the data of various authors on plaice, reef-fish, sunfish, desert pupfish, hatchery brook trout, inconnu, carp, *Leucaspius* and concluded that there was a general trend for the K line (gross efficiency) to fall with increasing rations. LeBrasseur (1969) confirmed this trend in experiments on juvenile chum salmon *Onchorhynchus keta* which received rations of fresh zooplankton of graded sizes. Moreover, mean growth rate of salmon on fixed rations, which ranged from 2·2-5·7% per day, was apparently independent of the size of the prey. The prey occupied three size groups—6-20 mm, 2·2-5·7 mm and ⩽1·5 mm in length—and consisted mainly of euphausiids and larger and smaller copepods respectively.

In summing up, then, from these experiemental assessments we can return to $T = \alpha W^{\gamma}$ (4.11) which gives an expression for describing the metabolic expenditure from the T line. The efficiency of utilization may be obtained, as we saw in (4.13), by

$$K = \frac{\Delta W}{R\Delta t} = e^{-a-bR}$$

or

$$\text{Log } K = \left(\frac{\Delta W}{R\Delta t}\right) = -a - bR$$

In the balanced energy equation

$$R = T + \frac{\Delta W}{\Delta t}$$

$$\frac{\Delta W}{\Delta t} = R e^{-a-bR}$$

By substitution, therefore

$$T = \alpha W^{\gamma} = R(1 - e^{-a-bR})$$

This system of equations has both theoretical and empirical bases, in which metabolism, body size, growth, rations, time etc. are "experimentally determinable variables reflecting various physiological and behavioural characteristics of particular life-history stages of species in given environmental situations" (Paloheimo and Dickie, 1966b).

Paloheimo and Dickie (1965, 1966a, b) claim that to measure food intake is a readier means of estimating metabolism than to measure oxygen uptake, because the former can be followed over extended periods whereas the latter is experimentally exacting, requiring confinement of fish in special apparatus tolerable only for a short time. However, the methods they employ also have shortcomings, because they depend on rather uncertain assumptions about food conversion, lack of standardization of holding conditions and the complexities of rescaling rations. There is even uncertainty, in feeding studies, about how much of a ration is actually consumed (Love, 1970). The work of Paloheimo and Dickie does, however, represent a very valuable effort to break away from assumptions about growth, based on a possibly unwarrantable transfer of concepts applicable to homeotherms, and attempts to establish some basic parameters of actual growth in fish and its relationship to metabolism. Much future work is likely to feature developments of the methods initiated by these workers.

Pandian (1967) reviewed the errors inherent in certain

assumptions about the energy equivalents of foods in growth experiments. He gave critical, detailed instructions for determining total nitrogen and total caloric equivalents of food consumed and gains in both nitrogen and calories made by fish during growth experiments. He also painstakingly estimated rates of feeding, digestion and absorption-efficiency for both total food and protein. For two species—*Megalops cyprinoides* and *Ophiocephalus striatus*—Pandian compared rates of conversion, for both protein of food and total food, by fish ranging in weight from a few grammes up to more than 120 g, with the published performances of sunfish *Lepomis* and the common carp *C. carpio*. Such studies will become increasingly valuable as man intensifies his search for animals with pronounced ability to convert the proteins of their food to a form suitable for human consumption.

Pandian (1967) also discovered that the absorption rate and conversion efficiency of food declined sharply in both *M. cyprinoides* and *O. striatus* ranging from less than 2 g to about 20 g. At larger sizes these rates declined very much more slowly, up to weights of more than 120 g. This finding also has a clear implication for fish culture and, of course, for our understanding of how fish of different ages utilize food.

Brett *et al.* (1969) recently investigated the effects of ration quality and temperature on growth of fingerling sockeye salmon *Oncorhynchus nerka*. The growing fish were maintained at temperatures of 1°, 5°, 10°, 15°, 20° and 24°C. The highest and lowest temperatures were about 1°C less and more than the upper and lower lethal temperatures, respectively, for this species. The rations comprised 0, 1·5, 3, 4·5 and 6% of dry body weight per day, plus one additional ration termed "excess".

Optimum growth rate for the two highest rations (6% and "excess") occurred at 15°C shifting to successively lower temperatures with each lower ration. At 1·5% the optimum growth occurred at 5°C. Among starved fish (ration = 0%) weight loss was maximal at about 15°–20°C. On rations of 1·5% fish failed to maintain body weight at temperatures in excess of 15°C and at temperatures above 21°C even those receiving rations of 3% lost weight. These results are depicted in Fig. 4.11. The derivation of various growth parameters using the data for salmon at 10°C is shown in Fig. 4.12. These figures bear an obvious kinship to Kleiber's (1961) scheme for mammal growth (Fig. 2.9).

Brett *et al.* (1969) also gave isopleth diagrams for gross and net efficiency, calculated according to the already cited formulas. What

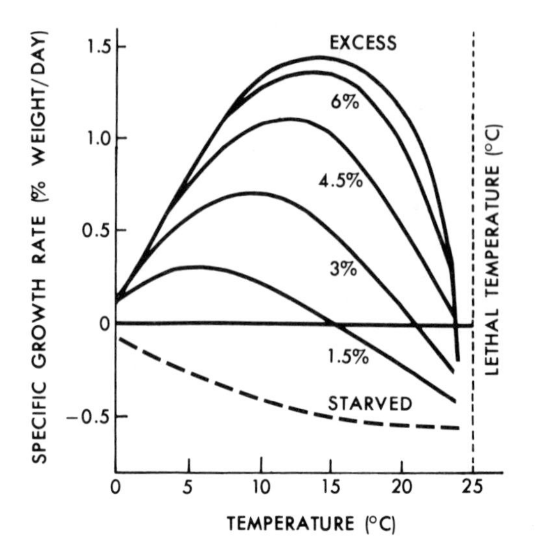

Fig. 4.11. Effect of reduced ration on relation between growth rate and temperature for 7 to 12 month old sockeye. The broken line for starved fish is a provisional interpretation. (Simplified after Brett *et al.*, 1969.)

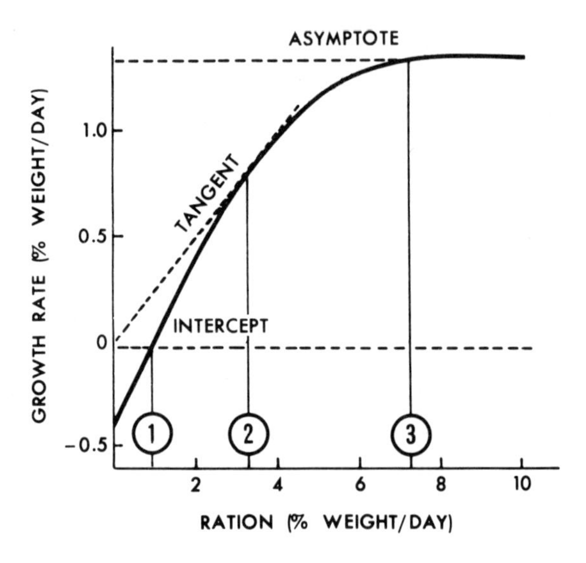

Fig. 4.12. Geometric derivation of various parameters of growth with accompanying ration, using data for fish at 10° C as an example. Encircled figures signify 1, maintenance; 2, optimum; 3, maximum. (After Brett *et al.*, 1969.)

these authors have termed the "favourable axis" has its centre in the same temperature range (9°–10° C) for both gross and net efficiency isopleths. For net efficiency, however, this isopleth axis is rotated anticlockwise to occupy a more vertical position with respect to the temperature axis (Figs 4.13 and 4.14). Whereas low temperature favours gross efficiency, the pattern for net efficiency shows that,

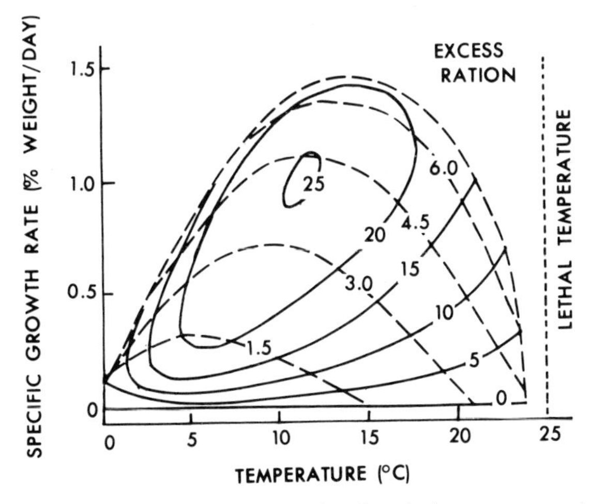

Fig. 4.13. Gross efficiency of food conversion in relation to temperature and ration, drawn as isopleths overlying the growth curves of Fig. 4.11. (After Brett *et al.*, 1969.)

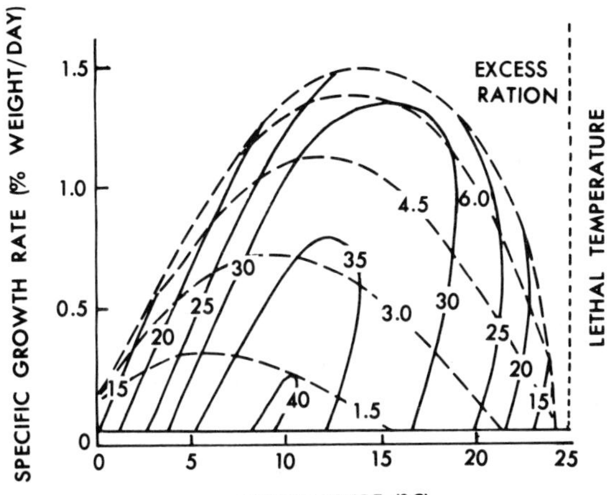

Fig. 4.14. Net efficiency of food conversion in relation to temperature and ration, drawn as isopleths overlying the growth curves of Fig. 4.11. (After Brett *et al.*, 1969.)

when the complex of energy requirements known as "maintenance" is separated from it, growth has a narrower and more restricted relation to temperature.

Paloheimo and Dickie (1966b) had suggested that the "K" line simply declined as rations increased, but Brett *et al.* (1969) were unable to confirm this trend. They found that, at all temperatures, K reached an optimum as rations increased, and suggested that the inference of Paloheimo and Dickie may have been based on cases in which the range of ration sizes considered lay above the optimum. Brett *et al.* also found that growth rate and fat and protein content of the body were directly related, whereas growth and body water were inversely related. This discovery may, as they have suggested, be very important to the ecologist attempting to gauge growth in wild populations on the basis of very little direct evidence.

Ursin (1967) recently published a highly mathematical treatment of fish growth in relation to respiration and temperature, which may prove to be valuable, but is extremely theoretical and relatively remote from the empirical basis of Paloheimo and Dickie's models. Understanding of Ursin's work is also complicated by an extensive mathematical notation which unfortunately seems to be incompletely explained.

B. Oxygen Uptake

Among poikilotherms, especially fish, basal metabolic rate or, as most fish physiologists have termed it in recent years, *standard metabolism* has been difficult to determine. Standard metabolism may be defined as the metabolism of minimal activity, the bare minimum, at a given temperature and excluding effects such as the specific dynamic (calorigenic) action (S.D.A.) of food for the energy cost of digestion and assimilation, that a normal living fish requires. Its value varies with size and species. Standard (basal) metabolism is determined most readily in man because activity may be controlled as desired and because human subjects can ignore the presence of minor environmental stimuli to which lower organisms unfailingly react. Basal metabolism in man is always measured in the post-absorptive state, i.e. sufficiently long after the last meal to avoid the effects of S.D.A.

Many studies on standard metabolism of fish have been vitiated because of extraneous stimuli affecting the activity level. Spoor (1946) indicated early the exacting conditions under which activity could be assumed minimal. He determined metabolism in goldfish

(*Carassius auratus*) by means of a paddle-shaped detector, correlating arbitrary "units" of activity with rate of oxygen uptake. The latter ranged from the value for standard metabolism, derived by extrapolating the oxygen uptake/activity curve to zero activity, to about four times higher, all under quiet conditions in an aquarium tank.

Beamish and Mookherjii (1964) described a respiration/activity chamber containing a mercury-filled thermoregulator, with built-in heater. A current controlled by the thermoregulator contacts operates the heater and keeps the temperature in the bulb of the thermoregulator slightly above that of the surrounding water of the activity chamber. When the contacts are properly adjusted in the chamber—closed and filled but without a fish—the heater is in circuit for 40-60% of the time. With a live and active fish in the chamber, turbulence currents created by its movement increase the heat loss from the thermoregulator bulb, which, in turn, increases the time the heater is in circuit. Increase in time of heating bears a close relation to uptake of oxygen in the sealed respiration/activity chamber over the same period.

Beamish and Dickie (1967) reviewed several other methods of monitoring activity in fish. However, as Beamish (1964) demonstrated, the problems in obtaining reliable data on standard metabolism can be formidable, particularly in view of the complex relation of its parameters to such environmental influences as oxygen tension. Fry (1957) demonstrated that for various fish species standard metabolism increases with temperature (Fig. 4.15). Warm water

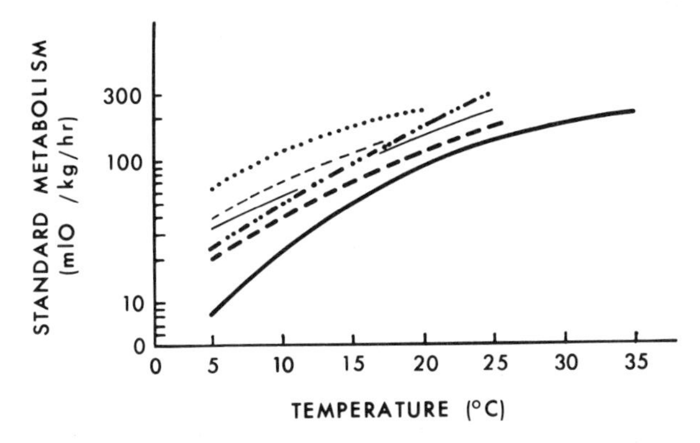

Fig. 4.15. Relation of standard metabolism and temperature in thermally acclimated fish, various sources; . . . . *Salmo*; – – – *Salvelinus*; –··–··– *Gobio*; —— *Squalus*; ▬ ▬ ▬ *Perca*; ▬▬ *Carassius*. All data from fish about 5 g in weight. (After Fry, 1957.)

species tend to have their peak oxygen consumption for standard metabolism at relatively high temperatures (35-40° C) whereas some cold water fish have their maximal oxygen demands at much lower temperatures, e.g. 15° C for brook trout and near 0° C for *Trematoma* spp., from the Antarctic (Beamish and Dickie, 1967).

Comparative physiologists once thought of the general metabolism of organisms as related to temperature in a manner closely comparable to that of chemical reaction rates, a relationship satisfied by the Van't Hoff equation

$$V = aK^T$$

There was also a widely held view that for biological actions, like many chemical actions, $Q_{10} = 2$, i.e. the rate of a process approximately doubles for every 10° C temperature rise. Both these relations hold in biology only as very rough approximations with no more than heuristic value at best. Approximate results based on unjustified analogy will no longer suffice now that we have many experimental devices with which to make precise measurements and if actual parameters of oxygen uptake and related variables such as temperature are required, careful experiments must be performed on living animals.

A sigmoidal curve might describe the relationship to temperature of processes such as growth more effectively than the Van't Hoff equation. However, as Needham (1964) has stated, the orthodox logistic would give only a velocity tending asymptotically towards a limiting maximum, whereas metabolism, growth and many other living processes have maxima at temperatures well below the upper lethal, their rate declining again with approach to the upper lethal (see section on "temperature and growth").

Figure 4.15 shows that Q_{10} of oxygen uptake decreases as temperature increases; at lower temperature ranges it may be about 2 or somewhat greater; and as temperatures approach the upper lethal it may decline below 2 or become negative.

If a fish is neither gaining nor losing weight and body composition is constant, standard metabolism may be regarded as an index of rate of utilization of food, as in mammals and birds; assuming that the relationship of oxygen uptake to food reserves used up is of the same order in fish as in mammals (see below). But standard metabolism curves for fish indicate one major difference between poikilotherms and homeotherms. In mammals, basal heat production is constant over a relatively narrow range of external temperature (Chapter 2, Fig. 2.9). Most mammals are to some extent temperature-independent and at low external temperatures their body heat production increases. This rising heat production can be obtained from an increasing food intake or, if there is insufficient food, by consuming stored materials such as fat, glycogen or even body fabric so that a new equilibrium is struck. But if the heat production continues at a high level in the absence of sufficient food, body reserves will be used up, body temperature will no longer be maintainable and the animal will become disabled by cold and die. Poikilotherms are not subject to the same hazards. If cooled too rapidly without opportunity to acclimate they may die. But a lowering of body temperature does not increase metabolic activity. On the contrary, body temperature and heat production follow changing external temperature and most temperate species (though probably not many tropical ones) readily survive prolonged exposure to temperatures close to freezing point if allowed time to acclimate. At such low temperatures metabolic rate is usually correspondingly low. Any scheme of the metabolic response of fish to changing temperatures intended as a "fish-equivalent" of that for homeotherms (Kleiber; Fig. 2.9) must allow for this great difference in thermal physiology. Furthermore, even Kleiber's scheme cannot apply directly to wild mammals or to captive mammals kept under simulated natural conditions. It applies only to animals incapable of modifying diet or life habits in the face of changing environmental temperature and deals with the essentially *automatic* responses of homeotherms, rather than the active responses that an animal may take to safeguard itself against temperature variations.

The activity of fish has proved difficult to measure and, perhaps significantly, investigators have found it hard to agree on appropriate terms in which to discuss activity (see Beamish and Dickie, 1967). Spoor (1946) and Beamish and Mookherjii (1964) showed how activity, rated in arbitrary units based on data from an activity recorder, gave a strong linear correlation with oxygen uptake. But there were no ready means for converting activity ratings to an objectively quantitative measurement of activity.

Starting with the work of Fry and Hart (1948) apparatus has been developed to investigate the metabolic cost of swimming. Originally, the apparatus consisted of an annular chamber which revolved at various selected rates. A fish swimming within the chamber was supposed to hold its position relative to a fixed point outside while the chamber and its contained water rotated. Results with goldfish *Carassius auratus* in this simple apparatus provided interesting preliminary data on the oxygen uptake—measured as the difference in the oxygen concentration of the water in the sealed chamber before and after a swim—associated with exercise. Fry (1957) and Wohlschlag (1961) described modifications of the original equipment. However, the apparatus has many serious shortcomings and is now mainly of historical interest.

Those investigators attempting detailed and extended study of the metabolic cost of swimming should consider building some sort of water tunnel—admittedly an expensive and fairly troublesome project though small models can be built at relatively low cost. A remarkable tunnel has been built and operated successfully by Brett (1963, 1964, 1965), who has performed a series of studies on swimming of sockeye salmon *Oncorhynchus nerka* and has written a highly readable narrative account of them (Brett, 1965). Other such systems have also been described (Blažka, Volf and Čepela, 1960; Kovalevskaya, 1956, as cited by Ivlev, 1964).

Before mentioning Brett's more important findings a note on water tunnels in general is in order. Brett (1963) has described how he surmounted difficulties in his work. To avoid major turbulence, non-uniform velocity etc. he installed wire mesh screens upstream of the test chamber paying particular attention to the design of the expansion and contraction cones leading into and out of it. Tests with dye, threads and small bubbles, together with study of fish behaviour in the tunnel enabled him to obtain a uniform flow profile. In fact, the water did not display laminar flow but "was put in uniform minute turbulence . . . completely free from massive turbulence, small eddies or vortices."

In such a tunnel, velocity is determined by means of pitot tubes and the fish swim freely in an accessible chamber, prevented from falling back by a wire screen bearing a slight electrical charge at the rear. There is virtually no lag in the chamber when flow rate is changed. The only serious objection is that the presence of the fish itself changes the flow characteristics of the water column it is swimming in; it does this in a way that is especially difficult to measure, because the drag on a fish-shaped object (e.g. a dead fish) is likely to be greater than on a live fish (Gray, 1957; Bainbridge, 1958, 1961). For instance, Brett (1963) observed freshly killed salmon suspended in the water stream in his tunnel, with air injected into the abdomen to give neutral buoyancy. Drag readings were taken, using dead fish secured to the end of a slim steel rod and at distances from the end. A tendency for the fish to oscillate when at the end of a length of thread was eliminated when it was attached directly to the end of the rod. Drag at the end of the rod, with and without the fish, was measured by means of a calibrated spring. Comparisons were made between the power which would be required for overcoming frictional drag (in the case of young sockeye salmon approximately 18 cm long and 50 g weight) and the power required for swimming over a range of velocities (Fig. 4.16). Brett (1963) determined that the maximum speed young sockeye (18·8 cm, 55·2 g) sustain more

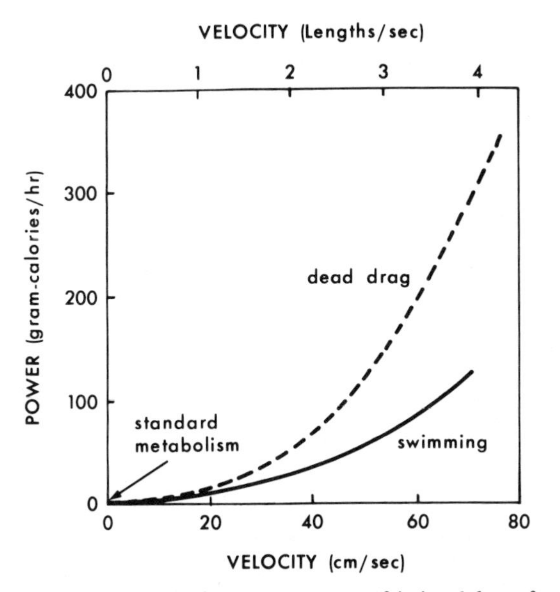

Fig. 4.16. Power required to swim by, or to overcome frictional drag of, a young sockeye salmon (18 cm, 50 g). (Simplified after Brett, 1963.)

or less indefinitely without going into cumulative oxygen debt as 4 body lengths per second (4 L/s), equivalent to about 71 cm/s. Figure 4.17 shows the ratio of the power which would be required to overcome drag on a dead salmon to that actually expended in swimming in the water tunnel by fish of similar size. The ratio reaches its highest value (2.2) at the highest velocity that can be maintained without cumulative oxygen debt. Brett (1963) regarded this as significant evidence that efficiency of swimming was greatest at this highest cruising speed (approximately 71 cm/s) which can be maintained indefinitely.

Figure 4.18 shows the metabolic rate of yearling sockeye swimming at various rates in the water tunnel. The standard metabolism was obtained on quite inactive fish. Figure 4.19 shows the effect of subtracting the standard metabolism for each

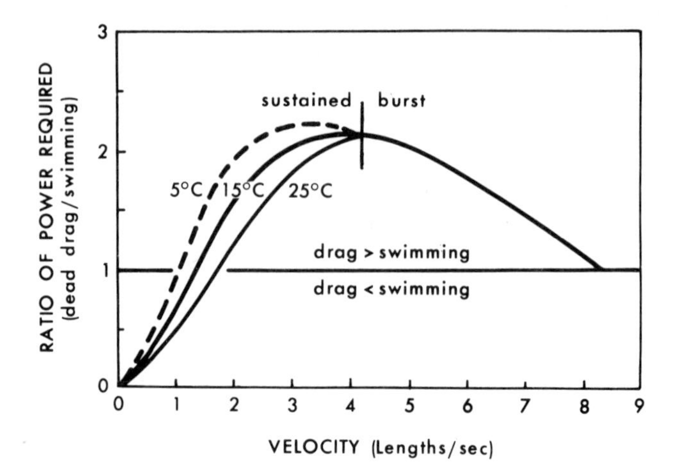

Fig. 4.17. Ratio of power needed to overcome dead drag to that expended by fish to swim at the same velocity, including sustained and burst speeds. Values above 4 L/s obtained from the ratio of experimental and extrapolated values. (Simplified after Brett, 1963.)

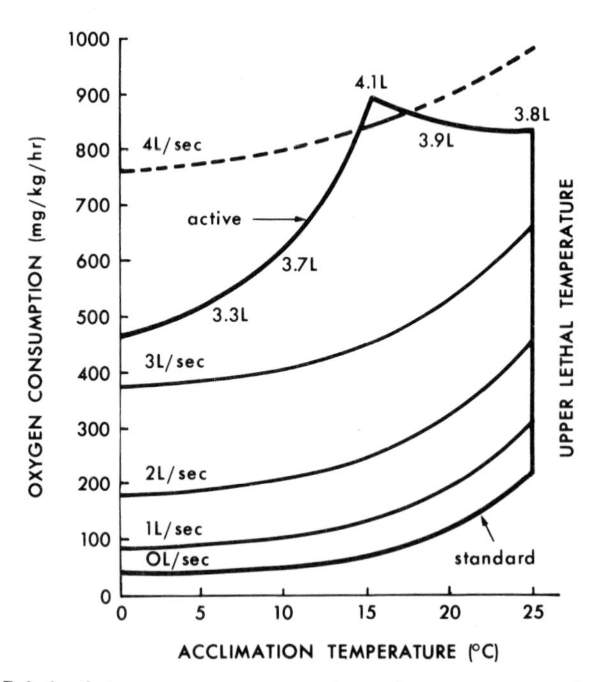

Fig. 4.18. Relation between oxygen consumption and temperature at various swimming speeds for yearling sockeye salmon (18 cm, 50 g). The broken line for 4 L/s is drawn in an area where rapid fatigue would occur since the speed at these temperatures demands a metabolic rate exceeding the active rate. (Simplified after Brett, 1963.)

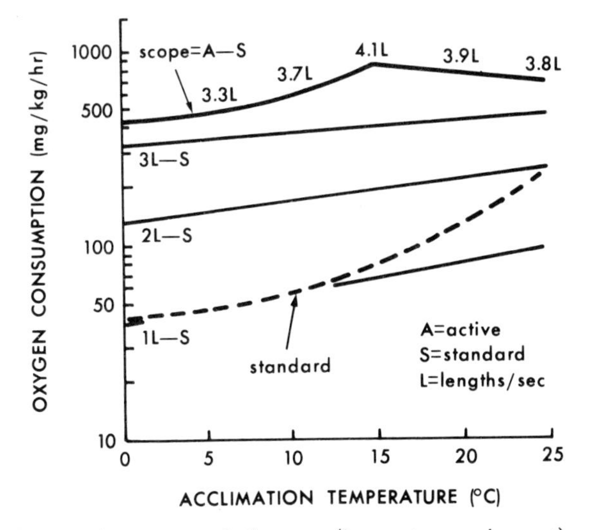

Fig. 4.19. Relation between metabolic scope (locomotor requirement) and acclimation temperature for various swimming speeds. Maximum scope with accompanying critical swimming speed is given in top line. The line for standard metabolic rate, subtracted in each case above, is shown for comparison to indicate "maintenance cost" at each temperature. (Simplified after Brett, 1963, 1964.)

temperature from the corresponding active rate, which gives *the scope for activity* at different rates of swimming and different temperatures—greatest at 15°C and 4 L/s for young sockeye. Furthermore, the minimal standard metabolism (at 5°C) was less than 5% that of the most active at 15°C, 4 L/s.

These studies by Brett have identified a group of important facts about the energy demands of activity in salmon and how these are met. As Brett (1965) stated, contrary to earlier beliefs about the incapacity of fish to meet the more vivid oxygen demands of exercise, the maximum performance of sockeye "is comparable to that of some mammals . . ."

It might at first appear reasonable to suppose that to study the power requirements of swimming would add appreciably to the methods available for investigating activity energetics of fish. It is, therefore, necessary to consider the findings of such studies.

Bainbridge (1961) reviewed various reports of the power required for swimming in fish or animals with a similar shape. A formula that permits the calculation of drag on a three-dimensional fish-like body is

$$\text{drag} = 1/2\rho A V^2 \, 1{\cdot}2Cf$$

where ρ is the density of water, A is the surface area, V is velocity, Cf is the coefficient of frictional drag. This formula is based on Gray's (1936) treatment for the dolphin, but in Bainbridge's adaptation a factor (1·2) is included to compensate for "form drag"—that due to displacement of water during forward movement and determined by the body form.

The power to propel a fish at velocity V is drag times velocity:

$$\text{Power necessary} = 1/2\rho A V^3 1{\cdot}2Cf$$

Bainbridge pointed out that

$$\text{Power available} = \frac{P\,.\,Wm}{2}$$

where P is a power factor which depends on unit weight of propulsive muscle and Wm is the weight of muscle. (The divisor, 2, is chosen on the supposition that only half the musculature can contract at a time.)

On apparently reasonable grounds Bainbridge assumed a swimming efficiency of about 75%, so that:

$$1/2A V^3 1{\cdot}2Cf = \tfrac{3}{4}\rho\,.\,\frac{Wm}{2}$$

which he simplified to

$$0{\cdot}64L^2V^3 Cf = P\,.\,Wm$$

if $A = 0{\cdot}4L^2$ as seems likely on empirical grounds.

The extremes at which this relation operate are set by values of Cf for laminar and turbulent flow. The values Bainbridge selected are related to the Reynolds number ($R = LV/\gamma$, where L is body length, V is velocity, γ is kinematic viscosity of water).

From this, he derived

$$V\text{Lam.} = \left[\frac{PWm}{0{\cdot}85L^{3/2}}\right]^{2/5} \quad \text{for laminar flow}$$

and

$$V\text{Turb.} = \left[\frac{PWm}{0{\cdot}01885L^{9/5}}\right]^{5/14} \quad \text{for turbulence.}$$

In attempting to compare observed velocities for fish of different sizes and muscle weights with calculations of velocity versus length, Bainbridge assumed that muscle comprised 50% of body weight, based on empirical evidence. He found that most observed velocities, whether of "burst" or "sustained" speeds, fell within the calculated range which lay between that under laminar and that under turbulent flow conditions—or else were less than the latter. The barracuda appeared to be the single exception, in having an unaccountably high velocity for its size.

Although we could hope to obtain approximate indications of power output of fish in nature when swimming at various speeds, if these could be estimated (see p. 118), it would be difficult to relate the energy of this to the total energy used in swimming. There will be a considerable difference between the physiological energy expenditure for a piece of activity and the actual generation of mechanical power needed to propel a fish's body at a given velocity. For this reason, it is probably simpler and safer merely to measure the total energy expenditure associated with a swim of a particular speed and duration.

C. Additional Commentary on Metabolism and Growth in Fish

In the earliest sections of this chapter and also in Chapter 2, it emerged that while growth curves (arithmetic, specific etc.) and derivatives of growth (such as condition) provide useful means for making gross and general comparisons, they do not lead to deep insights into the processes of growth.

Next, temperature and growth were considered and here it was shown that temperature must certainly have important effects and that the maximum growth rate often occurs many degrees below the upper lethal temperature. However, the relevant experimental studies were to some extent confounded by unsatisfactory feeding regimes, varying population density, non-standardization of food etc. In the observations on wild populations the difficulty was in deciding whether effects of temperature were direct (physiological) or indirect (through food). There seemed to be much room for critical laboratory and field experiments.

The next section dealt with "metabolism" as "food and growth" and the following section will consider the same topic in terms of "oxygen uptake". The total cost of living can be thought of as the sum of all the food needs for growth, maintenance, repair, activity, excretion etc. which Weatherley (1966) demonstrated in a very simplified scheme. There have been formidable operational difficulties in getting satisfactory data for these energy needs but some of the problems are now being solved. It is simple to perform respiratory experiments if the apparatus itself does not, through space restrictions, impose such deleterious conditions on fish that their metabolism is seriously affected, although even then there may be ways to monitor the extent of such effects and they may sometimes be prevented or at least allowed for. Since Spoor (1946) demonstrated that fish are unlikely to reach a state of standard metabolism unless carefully guarded from extraneous stimuli, many investigators have taken pains to insulate fish in their respiration chambers. The knowledge that even slight excitement may appreciably increase oxygen uptake can be put to use in allowing for such an effect when it does occur. Brett (1963) showed this when he drew a line of metabolism versus temperature that passed through the *lowest* points on his graph for sockeye salmon, claiming that the higher values were due to the fish being somewhat excited. Problems such as this are encountered also by those studying mammals and it was in mammals that the investigation of metabolism through oxygen uptake was largely developed. Investigators must certainly not ignore these problems when they arise in fish, but the entire method need not necessarily be in jeopardy. In suggesting that the method was seriously disadvantaged as compared to the "food intake" analysis of metabolism, Paloheimo and Dickie (1965, 1966a, b) and Beamish and Dickie (1967) have perhaps overstated the objections which are essentially technical ones.

A lack of really satisfactory chemical methods has been a major problem for investigators of fish metabolism through gas exchange. Titrimetric methods for measurement of oxygen and carbon dioxide dissolved in water have long been common in limnology and oceanography. Oxygen can be estimated fairly accurately and rapidly by such means. The recent advent of the oxygen electrode has made possible serial oxygen sampling on a semi-, or even completely, automatic basis. Carbon dioxide continues as a problem because titrimetric estimation, using an indicator such as methyl orange, is complicated by the frequently large concentrations of carbon dioxide in water and the relatively large error of the estimate. Carbon dioxide can, however, be accurately determined by direct extraction in a Van Slyke apparatus.

It would be highly desirable to be able to determine with precision the respiratory quotient (*R.Q.*) in fish. If *R.Q.* is known (and making an approximate allowance for nitrogen excretion), from a knowledge of the calorific values of different foods it should be possible as in mammals to derive with reasonable precision the energy equivalent of the oxygen consumed in terms of food utilized.

Kutty (1968) investigated *R.Q.* in goldfish and rainbow trout over a wide range of activities. He used a modified Fry (annular chamber) respirometer, in which spontaneous swimming activity was recorded as the number of times the contained fish interrupts beams of light falling on photocells. Kutty used a phospate buffer solution instead of water to prevent major pH changes due to released CO_2. Oxygen was determined by means of a dropping mercury electrode, carbon dioxide by the Van Slyke method. At near standard metabolism (activity minimal) the *R.Q.* was slightly less than unity (i.e. similar to the mammalian value), increasing to considerably more than unity at high levels of spontaneous activity, even exceeding a value of 2. There was considerable variation in the value of the *R.Q.* at any activity level when repeatedly determined on single fish. Kutty also found that, as ambient oxygen concentration decreased from near saturation, spontaneous activity in both goldfish and rainbow trout would first increase, then either decline somewhat or remain steady as ambient oxygen reached 50% of air saturation. Accompanying these changes in activity the oxygen uptake and carbon dioxide production increased, but to the same extent, so that *R.Q.* remained constant down to the critical oxygen value of 50%. Below this, both oxygen uptake and carbon dioxide output decreased, but the former more rapidly than the latter, so that the *R.Q.* rose.

In goldfish forced to swim (in a Blažka tunnel respirometer), Kutty found that oxygen consumption rose with rate of swimming until a speed of 400 body lengths per hour was reached, when a rather abrupt decline to a lower level of consumption indicated that the fish can shift to a more sparing use of oxygen at higher sustained swimming speeds. Smit (1965) found much the same in experiments on goldfish. Kutty also found that oxygen uptake and carbon dioxide output showed similar declines during the swimming of goldfish at low steady rates for sustained periods (73 h).

These sorts of studies do make it obvious that any simple assumptions about the energy metabolism of fish from their respiratory exchanges would be unwise. On the other hand, as work on mammals (especially man, e.g. Cureton, 1951; Johnson, 1960; Chapman and Mitchell, 1965; Karpovich, 1965; Morehouse and Miller, 1967) has shown, the complexities of relating the respiratory metabolism of exercise to food and energy metabolism have not deterred researchers in this area nor should they deter fish workers.

The calculation of nitrogen excretion of fish is performed in the same way as for other animals, except that urine itself is not usually collected but a sample of water in which the fish have spent a known time—in a small container a few hours may suffice. Karzinkin and

Krivobok (1964) have described the methods and have also given detailed directions for the calculation of nitrogen balance in fish (see also Pandian, 1967). However, there are confounding difficulties in interpreting the meaning of nitrogen values thus obtained because of the unpreventable accumulation of faeces in tanks.

Brett's (1963, 1964, 1965) work is an interesting attempt to apply respiratory metabolic studies to interpretation of problems of mass movement and survival in sockeye salmon. The present relevance of his work, which actually concerns the salmon spawning migration, is that it involves a period of "degrowth" when the salmon begins a journey with certain food reserves laid down in the tissues, the loss of which (a result of swimming activity and of other general catabolism) is not compensated by food intake because spawning salmon do not feed. Brett described salmon swimming in cages in moving river water, in which body fat and protein were determined following a swim of 12·7 days at 3·9 km/h (equals 896 kilometres). Fat and protein in these were compared with fat and protein in control fish killed before the long swim began. Energy consumption† compared fairly well with data from short-term swims in the water tunnel respirometer.

Idler and Clemens (1959) followed the spawning movement of a wave of migrating sockeye salmon up the Fraser River—1024 kilometres in 20 days. Had they breasted the mean river velocity of 4·8 km/h, the fish would have averaged 7·0 km/h. Respirometer tests indicate that sockeye cannot sustain swimming at this speed, a cumulative oxygen debt soon forces them to quit. It can therefore be inferred that sockeye skilfully follow minimum flow paths in the river to progress as they do. This could not have been foretold from field observations, even if accompanied by estimates of utilization of food reserves. Experimental investigation of the cost of swimming and ability to maintain speed was also required. Furthermore, during the migratory swim the fish consumed about 90% of body fat and about 15% of body protein. In fact, they spent energy at 80% of the maximum maintainable rate—the equivalent of sustained swimming at 4·2 km/h. Their skill in choosing a path apparently allowed them to reduce the speed they would have required, had they swum directly into the average current, by about 2·7 km/h. In the sea the signs are that sockeye migrate shorewards at about 1·8 km/h, which is near their most "efficient" speed.

† The energy equivalents for food consumed, assumed by Brett, were those established for warm-blooded animals. Further study may allow corrections to be made for foods eaten by fish.

These findings are recounted because of their implications for the investigation of fish growth. Brett and his colleagues have, by the study of metabolism during one life stage of salmon, opened the way for work on a variety of connected problems. For instance, what is the maximum journey a salmon may undertake, given certain food reserves at the beginning, assuming water temperature and velocity can be determined with reasonable accuracy? Such a computation—which may now be successfully performed—is of great potential practical importance in managing water-ways for salmon (or other species).

When attempting to compute the metabolic cost of the swimming of stenohaline or migratory fish one may have to take into consideration the energy requirements of osmoregulation. According to Farmer and Beamish (1969), assuming energy required for osmoregulation in the euryhaline fish *Tilapia nilotica* was zero at the isosmotic salinity of 11·6 approximately 29% of total oxygen consumption was required for osmoregulation at 30‰ when swimming at 30-50 cm/s and 19% was required at 0, 7·5 and 22·5‰. Rao (1968) found a similar order of values for active rainbow trout at different salinities.

From the papers of Paloheimo and Dickie (1965, 1966a, b) Warren and Davis (1967), Beamish and Dickie (1967), Mann (1967) and Weatherley (1966) certain general considerations of the partitioning of energy into growth, activity etc. may be derived. By using procedures mentioned in these reports it should eventually become possible to construct a scheme to show the fate of food (or its energy equivalents) and hence to be able to make comparisons between individual fish or between populations of fish. It should also become possible (and see Weatherley, 1966) to obtain enough actual values for these various aspects of energy utilization for better understanding of the differences in growth between populations, year groups etc. and hence to comprehend the population dynamics of fish in a more revealing light.

Measurement of activity among wild fish is admittedly a taxing problem. In river-migratory species, such as salmon, the problem is at least reduced to a one-way, sustained movement during which no food is taken. But most fish undertake no regular migrations and it is their activity while they are feeding and growing that we want to know about, rather than activities while they are not feeding and are degrowing. Simply to mark and release fish provides no useful index of activity, because on their recapture we are entitled to assume only that they have swum in a straight line at a uniform speed since release, which is very unlikely even when the time between marking and recapture is very short.

It may soon become practicable, as it is already for many terrestrial animals, to embed transistorized sensors within the fish to transmit information on body temperature. If the temperature changes that accompany swimming in a water tunnel at a given speed were known and could be correlated with oxygen uptake, oxygen uptake of fish in the wild could be inferred on the basis of their transmitted body temperatures (at given water temperatures) (see Stevens and Fry, 1970).

Meanwhile, important work on fish activity in nature has recently been performed by Hergenrader and Hasler (1967) who observed the swimming rates of yellow perch *Perca flavescens* in Lake Mendota, Wisconsin over a period of a year, using sonar equipment. They were able to determine roughly the size of schools and to separate patterns made by single fish and they found a strong correlation of average swimming speed with prevailing water temperature, with a linear Q_{10} relationship (from 10° to 25°C) of 1·5. Hergenrader and Hasler found that there was a considerable range of swimming speeds at different temperature ranges, so that total energy cost of swimming would probably have to be computed from the frequency distributions of speed rather than means (Fig. 4.20), because at least three of these distributions are strongly skewed. These authors have

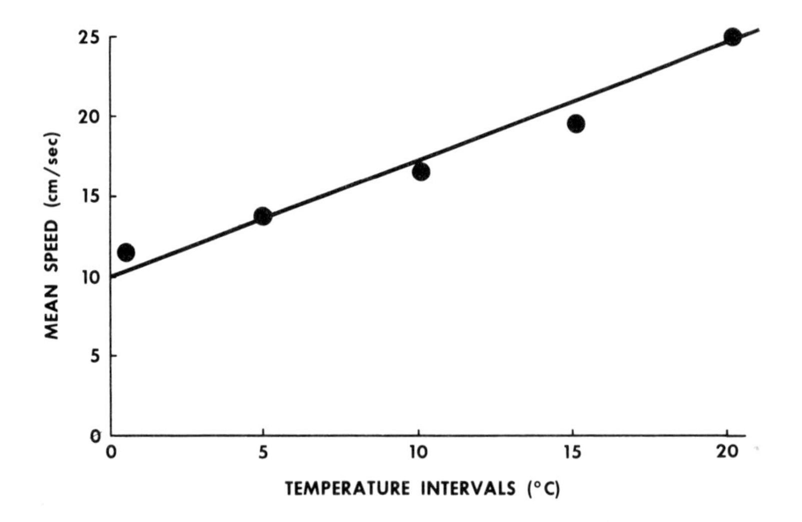

Fig. 4.20. Mean swimming speed of schools of yellow perch at different temperatures. Number of observations represented by each point ranged from 270 to 536. (After Hergenrader and Hasler, 1967.)

made a successful beginning on this problem. Their reasons for tackling it are the same as those at present under discussion: "By knowing not only how rapidly fishes move, but also the duration of activity, estimates of this part of the energy balance sheet can be calculated" (Hergenrader and Hasler, 1967, p. 373). Henderson *et al.* (1966) have reported some success in tracking individual active fish in Lake Mendota using a small ultrasonic transmitter in the form of a tag.

In a complete study of a population such as the Lake Mendota perch, full data on the diel activity cycle and the sizes of fish in the schools would be needed, for these factors would determine how much energy the schooling fish require to swim at a certain speed. In combination with study of the actual metabolic cost of swimming at various speeds this approach should eventually yield a full statement of the energy requirements of an active population in nature.

In concluding this section a few points remain for mention. Davies (1963, 1964), in attempting to construct a total energy balance sheet for goldfish, concluded that the energy extraction efficiency from the food increased with rations up to a certain level depending on temperature (about 5 mg dry wt/g fish/day at 12° C and about 12 mg dry wt/g fish/day at 21·5° C)—the reverse of what Paloheimo and Dickie (1966b) and Beamish and Dickie (1967) have reported. However, the explanation of these results may perhaps be similar to that of Brett *et al.* (1969) who, as already mentioned, found that there was an optimum efficiency of a certain ration size, with greater and less efficiency above and below that ration. Davies also suggested (1966) that heat production was inversely related to degree of crowding and (1967) that net efficiency of growth (see also Brown, M. E., 1957) was higher when food intake just exceeded maintenance requirements. Unfortunately, the peculiar conditions of many of Davies' experiments make the latter studies (1966, 1967) difficult to appraise, particularly as they do not allow for the possibly very important requirement of energy for activity in wild populations.

Gerking (1966) considered growth of bluegill sunfish *Lepomis macrochirus* in eight lakes, which he carefully chose because the populations exhibited widely varying growth rates. Frequent resampling of the populations showed that the different annual increments in size were a product not of differing specific growth rates, but of length of growing season. Thus Fig. 4.21 shows that average steepness of curves of growth for several age groups in two lakes differed little, but that in the one with the greater annual increments in size, the growing season was about 50% longer than in

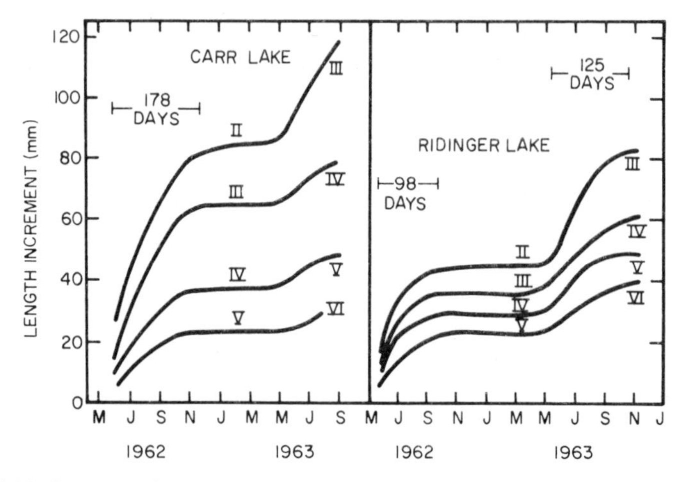

Fig. 4.21. Seasonal gains in length in two bluegill populations exhibiting growing seasons of different lengths. The abscissa is divided into monthly intervals beginning with May. The ordinate is the length gain of various age groups at several collecting dates beginning in the spring of 1962. (After Gerking, 1966).

the other. Group III fish from several lakes (Fig. 4.22) had size increments during the first month of the growing season of 12·6, 12·9 and 13·9 g respectively for populations with rapid, intermediate and slow annual growth rates. The Ridinger Lake population had the shortest growing season and the poorest annual growth, yet it also showed the greatest increment during the first month. Clearly, temperature is not governing growth in any simple fashion here. Gerking suggested that rate of temperature change, light regime and food may combine to influence secretion of growth hormone.

It will be useful to recall briefly the various sorts of investigations of growth that have been mentioned.

Examination of a growth curve, whether empirically fitted or the product of a biomathematical analysis, can in itself offer little indication of the forces and materials which have influenced its shape. For that it is necessary to assess the amount (or energy equivalent) of food a fish is ingesting and its channelling according to schemes such as that of Needham (Chapter 2) or Warren and Davis (1967); see also Davies (1964). The analyses of Paloheimo and Dickie (1965, 1966a, b) show that temperature, possibly light, changing food regimes and activity patterns are all likely to influence growth.

A weakness of all these studies is that, although sound in principle and in general even in the quantitative data in energy balance sheets (e.g. Davies, 1967), they are, like textbook diagrams of simplified

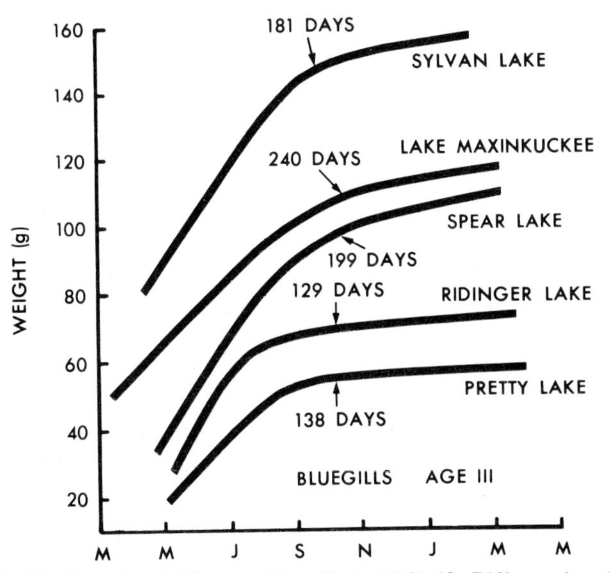

Fig. 4.22. Weight gain of III group bluegills in 1962-63. Different lengths of growing season in different lakes are indicated. The five populations represent rapid (Sylvan and Maxinkukee), intermediate (Spear) and slow (Ridinger and Pretty) growth rates. (Simplified after Gerking, 1966.)

nitrogen cycles, somewhat deficient in quantified detail. As work on this problem proceeds the essential question will be how to obtain quantification. Analytical and experimental studies could be pursued endlessly and plans of what may be happening in wild fish populations inferred from these. But perhaps a slightly more heterodox and immediate approach can be obtained by retaining the ground plan of accepted principles, while taking the tools and methods of the laboratory into the field. Then wild populations may be studied at first hand. After all, in putting physiological and mathematical models to work in ecology it is necessary to consider simultaneously age-specific mortality rates, recruitment, competition, predation, length of growing season and growth itself. This programme will be developed, mainly in terms of systems ecology, in Chapters 9 and 10.

The one basically reliable approach is to construct detailed energy balance sheets for *populations,* in which recruitment, death, emigration, activity and growth, as well as food wastage and excretion are all accounted for quantitatively. Some of the means for this are now available and others are being developed, so that by a judicious mixture of direct field sampling and selected laboratory experiments good results ought to be obtainable. The experimental parameters should be determined by the measured field conditions, which would then provide guide lines outside which it would normally be pointless for the ecologically-oriented experimenter to stray.

If detailed balance sheets could be constructed so as to bear an immediate relation to several actual populations of a species, in which growth rate differed widely, it might be possible to determine how and to what extent the energy characteristics of the populations govern the growth patterns of their members.

V. PHYSIOLOGICAL AND CELLULAR ANALYSES

A. Cells and Growth

The latter part of Chapter 2 traced the contribution of cells to growth in vertebrates. Its generalizations were derived from observations on only a small variety of vertebrate animals—those that have been used most commonly by experimental biologists. Fish have scarcely been studied at levels suitable to contribute to this field of biological knowledge. From articles such as that by Love (1957), a few very tentative speculations on the cellular aspects of growth in fish can be put forward. Fish muscles are deficient in connective tissue proteins in comparison with, for instance, bovine muscles. Probably this indicates that the metabolically most active proportion of cells in fish muscles is relatively high. On the other hand, if fish muscle cells are part of the "static cell populations" of the body (Goss, 1964), as in those vertebrates that have been investigated, then repair of damaged fish muscles, with their deficiency of connective tissue, may be a rather slow process. Connective tissue, part of the "expanding cell populations" of the body, also heals more rapidly than damaged muscle. A critical comparison of repair of muscular lesions in fish and other vertebrates—particularly mammals—might be illuminating.

In vitro culture of fish tissues was begun nearly 60 years ago when Oskowi maintained explants from embryonic trout for 24 h in Ringer's solution and frog lymph. In their review of fish cell and tissue culture Wolf and Quimby (1969) have explained that relatively little work was done on fish during the succeeding 30-40 years. From the late 1940's reports began to appear more frequently, though initially much of the work had a bias towards pathology—e.g.

cultivation of tumour tissue and viruses. Subsequent work has been mainly concerned with the elaboration of optimal media and the establishment of cell lines from teleosts, with only a minor effort on elasmobranchs or cyclostomes. A significant discovery was that many teleost tissues can be grown as monolayers in mammalian types of culture media with no more than minor modifications. A number of media designed specifically for fish tissues have, however, been elaborated and these, together with methods for preparing fish tissue for culture, have been described by Wolf and Quimby (1969).

Tissues of fish including embryos, gonads, trunk muscles and fins have been cultured in established cell lines. And we may note that fish cells are useful material to the comparative physiologist, because their metabolic rate can be more fully manipulated through temperature than that of mammal or avian tissue. It is also an interesting fact that, among the established animal cell lines, only those from mammals and birds are more numerous than those from fish (Wolf and Quimby, 1969).

Love (1970) has reported that in cod *Gadus morhua* an increase in body length from 30 to 100 cm is accompanied by an increase in diameter of muscle cells, though not in their number. However, Greer-Walker (1971) described an increase in red and white axial muscle fibres in cod, with a steep rise from less than 10 000 at fish lengths less than 10 cm to about 65 000 at 40 cm, followed by a slower rise to 115 000 at 120 cm. He also found that as fish grew from 0 to 120 cm in length, muscle fibre diameter increased to a maximum at 80 cm for white fibres and to an asymptote at the same length in the case of red fibres. Greer-Walker also showed that red fibres come to occupy a greater proportion of the total cross-sectional area of muscles in cod as the fish become larger. This is of interest with respect to swimming activity, because cod of more than 90 cm have white fibres which are more numerous than at smaller fish lengths though occupying a smaller proportion of the total cross-sectional area of the muscle. Greer-Walker has noted that in larger cod this may be an adaptation which functions to offset internal friction in swimming movements which is presumed to become greater as fish increase in size.

Clearly, there are important questions here which should soon yield interesting answers to those investigating both growth and swimming activity.

It is to be hoped that work on fish tissues will soon enter a more analytical phase so that the sorts of experiments that have revealed so much of the functional organization of cells in mammals and birds will also be performed using fish cell lines. A considerable advance in the understanding of fish growth could then be anticipated.

B. Regulation and Integration of Growth

Protein requirements of fish are, according to Love (1970) considerably higher than those of homeotherms, though one would prefer more data on this, especially from herbivorous fish. The list of essential amino acids for fish is the same as that for other vertebrates (Chapter 2, p. 42). As fish increase in age and size and become sexually mature there is evidence of change in the proportions of the amino acids among the tissues of the body. Thus, in *Oncorhynchus keta,* proline and glycine—of which there are high concentrations in the collagen of connective tissue—decrease in gonads of both sexes during maturation. It has been inferred that this decrease is due to an increase in bulk of ova and sperm, giving a reduced proportion of connective tissue in gonads. Accompanying these changes, lysine, histidine and arginine increase in concentration in the male gonad and leucine and isoleucine increase in both sexes. The glycine/alanine ratio has been used as a maturity index, ranging from about 1·3 (in immaturity) to about 0·5.

As a result of the spawning migration in salmon, when the body tissues are being used for energy, the muscles may show increased proportions of proline and glycine, because it is the more readily mobilized amino acids that are utilized rather than those of the collagen of connective tissue (Love, 1970). Questions of protein synthesis or neoformation and related problems (Pandian, 1967) are poorly understood in fish, though their general features will perhaps not depart greatly from those in other vertebrates.

That the role of proteins in the body's maintenance is a complex one is readily seen in data from common carp *Cyprinus carpio* starved for eight months (Creac'h and Cournede, 1965; Creac'h, 1966). Total free amino acids of muscle fell to about 16% of normal and glycine and histidine were especially depleted. Most free amino acids decreased in the intestine but many of those of the liver, kidney and spleen increased, especially leucine. Physiological understanding of these changes is meagre. During starvation, water content of fish tissues can increase from about 80% to the order of 90-96% (Love, 1970). While this is going on, serum proteins (in carp) can drop from about 4% to about 3%, the albumins being diminished first then the α- and β-globulins. Muscle protein falls considerably, in cod, during several months of starvation. At the whole organ or tissue level, brain and heart are highly resistant to starvation, the tissues most affected being liver, kidney, spleen, intestine, muscle—in that order. The brain lipid fails to decrease even when fish are about to die from starvation and when liver lipid has decreased from 40 to 2% of weight (Love, 1970).

As cod grow from 30 to 100 cm their muscle DNA apparently decreases in concentration. Love (1970) thought this might be partly due to a decrease in number of cells per unit of body weight as fish increase in size. The decrease was not great enough to be explained entirely in this way which suggested that some DNA synthesis could be occurring during growth. At present, however, it is difficult to reconcile such explanations with Greer-Walker's (1971) study on cell growth in cod mentioned above.

Complex changes occur in water and cation balance of blood and tissues of fish during growth, development and maturation, changes in sodium being especially notable (Love, 1970).

Vitamins have received little study in relation to fish growth. However, Kitamura *et al.* (1967) were able to characterize poor growth as among the symptoms, in rainbow trout, of deficiency of vitamins A (retinol, axerophthol), B (aneurin, thiamine), B2 (riboflavin), B6 (pyrodoxine) and pantothenic acid. Vitamin C (ascorbic acid) deficiency is associated with deformity of vertebrae and operculars. As is usual with vitamin deficiencies, a wide range of other symptoms are associated with the symptoms of poor growth. The list of symptoms may be found in Kitamura *et al.* (1967) or as Table XXXII in Love (1970).

The following notes on growth hormone in fish draw freely on the monograph of Pickford and Atz (1957) and on Ball, J. N.'s (1969) review. Growth ceases as a result of pituitary extirpation in elasmobranchs and teleosts. Pickford and Atz described typical changes in hypophysectomized teleosts, which are associated with poor appetite. These changes may be a slow weight loss leading to emaciation or oedema resulting from blocked kidney ducts. Some hypophysectomized fish may gain weight from accumulating food reserves, when livers are laden with fat and glycogen as in the "winter period of sexual quiescence." In many instances remission of these symptoms and resumption of growth follows administration of pituitary extracts. It appears that even though hypophysectomized fish will continue to eat and remain in fairly good health they cannot grow until they receive pituitary extract containing growth hormone. Swift and Pickford (1965) used hypophysectomized *Fundulus heteroclitus* to assay the annual cycle of growth hormone content in the pituitary of *P. fluviatilis* in Windermere. In winter–a time of no growth for perch–the concentration of growth hormone in the pituitary was correspondingly low as judged by its effects on growth in *F. heteroclitus*. Through spring, the content of growth hormone increased to a maximum in June, declining again in August. This cycle of stored hormone corresponded closely with the observed growth cycle in perch.

There is a critical temperature below which the growth hormone is relatively unsuccessful in restoring growth in hypophysectomized fish but above which growth may resume. Pickford and Atz (1957) found that temperature influenced the response of hypophysectomized *Fundulus* to growth hormone. This does not mean that above a critical temperature there is a significant correlation between temperature and growth. Temperature evidently acts merely as a trigger (see again earlier section on Temperature and Growth). However, it does appear that it may be via the rate of secretion of growth hormone that temperature exercises its effects on the growth cycle of fish under natural conditions in varying climates.

As indicated in Chapter 2, fish respond to bovine growth hormone, though mammals do not respond to fish growth hormone (Pickford and Atz, 1957; Ball, J. N., 1969). The chemical constitution of growth hormone is unclear, though Wilhelmi (1955)

Table 4.1

Comparison of the physico-chemical properties of beef and fish growth hormone.
(Based on data of Wilhelmi, 1955. Modified slightly after Pickford and Atz, 1957)

Physico-chemical Property	Crystalline Beef Growth Hormone	Crystalline Fish Growth Hormone
% α amino nitrogen (alanine and phenylalanine	0·03	0·025
% amino nitrogen (lysine)	0·6	0·7
% amide nitrogen	0·7	0·65
Isoelectric point	6·95	6·2–7·2
Sedimentation constant ($S_{20} \times 10^{13}$ cor.) at pH 9·7, 0·6–1% protein	2·72	1·78
Corresponding molecular weight	44–47,000 (est.)	22–26,000 (est.)

considered the properties of a crystalline preparation of growth hormone from hake pituitaries. Table 4.1 compares crystalline beef hormone and fish growth hormone. The proportions of constituents appear similar, but the molecular weight of beef growth hormone is about twice that of fish growth hormone; but see also Ball, J. N. (1969).

While the relation of growth hormone to other hormones is still obscure there is evidence of a synergism with thyrotropin (which by itself lacks growth-promoting ability). Whether thyroxine has a positive effect on growth in fish is still an unresolved issue despite nearly 20 years of research by a number of investigators. On the other hand, it appears fairly certain that thyroxine does play a part in forming the skeleton of the developing fish (Gorbman, 1969).

C. Inheritance of Growth Rate Capability

There are indications, usually indirect, that growth rates of different populations of fish constituting separate genetic pools may often differ in capacity for growth and hybrid vigour, exemplified in superior growth as has sometimes been claimed. Some relevant experiments are summarized by Brown, M. E. (1957) and Frost and Brown (1967). It appears, however, that most such work is vitiated by failure to allow for such complications as the intensity of competition for food (Chapter 5). Since many fish have demonstrably flexible growth it is necessary to ensure, if the growth of one group of fish is to be validly compared with another, that factors such as population density, food, size/age composition and relative activity levels are carefully standardized. Unfortunately, it appears that, in investigating the significance of reputed genetic influences on growth, insufficient attention has been given to these matters.

This, of course, is not intended to mean that there are no growth differences that have a genetic basis—in one sense all must have such a basis. A sex-linked difference between the growth rates of males and females in one population is a frequently encountered demonstration that genetic endowment influences growth (Le Cren, 1958). Furthermore, many "small" fish species never reach the ordinary size-for-age of "large" species even under the most favourable conditions for growth. On the other hand, as Fig. 4.23 indicates, individuals of "small" species may, at their maximum rate, sometimes grow as rapidly as the slowest growing individuals of species that are normally regarded as "large" by comparison. However, until meticulous experiments—which will have to be

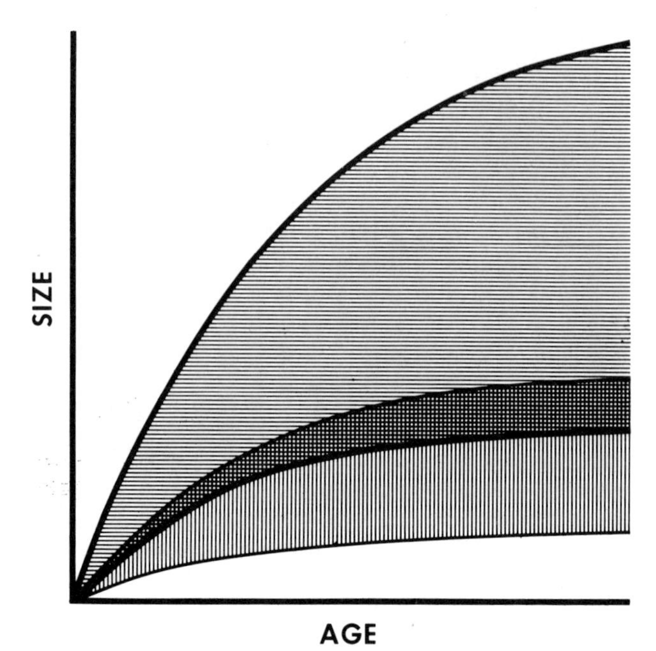

Fig. 4.23. Comparison of growth rates of a rapidly growing fish species and a slowly growing one. The overlap between shaded areas shows coincidence of poor growth of the former species with good growth of the latter. The upper of each set of curves represents the maximum potential growth rate for each species. (After Weatherley, 1966.)

carefully designed, with a full appreciation of possible sources of confusion—are performed, it will remain difficult to prove in most cases that fish inherit more than an upper limit to the size which they may attain.

5 | Food, Competition and the Niche

The processes of growth in fish and other animals and methods of determining their course in wild populations have been discussed in Chapters 2 to 4. In this and subsequent chapters growth in relation to the fish population will be the topic.

I. FEEDING HABITS

Digestion and absorption of food are necessary conditions of growth and since competition and growth have meaning for ecologists largely in relation to food, we begin by considering different aspects of fish feeding relationships. A formula for competition has been proposed in this chapter, one of the terms of which, M_p stands for rate of production of food organisms available to fish. We begin by examining what is meant by *available food.*

Many studies of production and energetics of ecosystems have been conducted along rather generalized lines (see references in Chapter 1). Detailed investigations of the production of food available to fish should be based on the concepts of these general studies but must fulfil rather more specialized requirements. In the first place, they should determine precisely which organisms fish consume and this calls for a dietary survey, although this alone would not serve to identify all available foods consumed by a population of fish, unless done with uncommon completeness. Many fish species have wide, variable food habits (e.g. Hartley, 1948) and though they may appear to show preferences, these often seem to be conditioned by the predominant food items among those available (Neill, 1938; Allen, 1941). Ideally, investigation of the food habits of a particular fish population would include detailed examination of its diet, critical feeding experiments in which foods apparently unconsumed in the field were offered alone and in measured combination with other foods, together with study of the published information on the food of the species; Weatherley (1959) attempted to meet some of these requirements for tench *Tinca tinca*. If possible, investigations of food habits should be extended ones; ecologists are often surprised by sporadic occurrences in the diet of a species of items generally thought to be unavailable or ignored. Moreover, a complete investigation would involve examination of the functional morphology and physiology of the alimentary system (see Barrington, 1957). While it is always difficult to be sure about what a population *might* consume under all circumstances, one can often make reasonable inferences about the possible diet—for instance, organisms which pass most of their lives deep in mud or under the surface of stones will be unavailable to fish, whose feeding methods exclude mud-sucking or which are too large to search under stones.

In some instances it may be relatively simple to infer the general nature of a fish's food from knowledge of its functional morphology. For instance, most strongly herbivorous fish have specialized alimentary systems that, from appearance alone, indicate their food habits. Many herbivorous fish are obligatory plant feeders, unable to ingest materials other than plants and lacking the alimentary system to digest them. The same principle holds for markedly carnivorous fish which also have highly specialized alimentary systems. Das and Moitra (1956a, b) described beautifully distinct examples of both such specialized types of alimentary system among Indian fishes of the Uttar Pradesh (Fig. 5.1). They found that:

(i) Herbivorous fish lack teeth, but possess fine gill-rakers that can sieve microscopic plants from the water; they also lack a true stomach (i.e. a highly muscular, acid-secreting, anterior distension of the alimentary canal) but possess a long, thin-walled intestine.

(ii) Carnivorous fish have teeth well-developed to seize, hold and tear and gill-rakers modified to grasp, retain, rasp and crush prey. There is a true flask-like stomach and a short intestine, elastic and thick-walled.

(iii) "Omnivorous" fish have alimentary systems of forms more or less intermediate between those of the extreme herbivores and carnivores, but there is a range of types between the extremes. The alimentary canal is much larger, relative to body length, among herbivorous than carnivorous fish, with the "omnivores" again occupying an intermediate position.

Nikolskii (1963) gave parallel descriptions of feeding specializa-

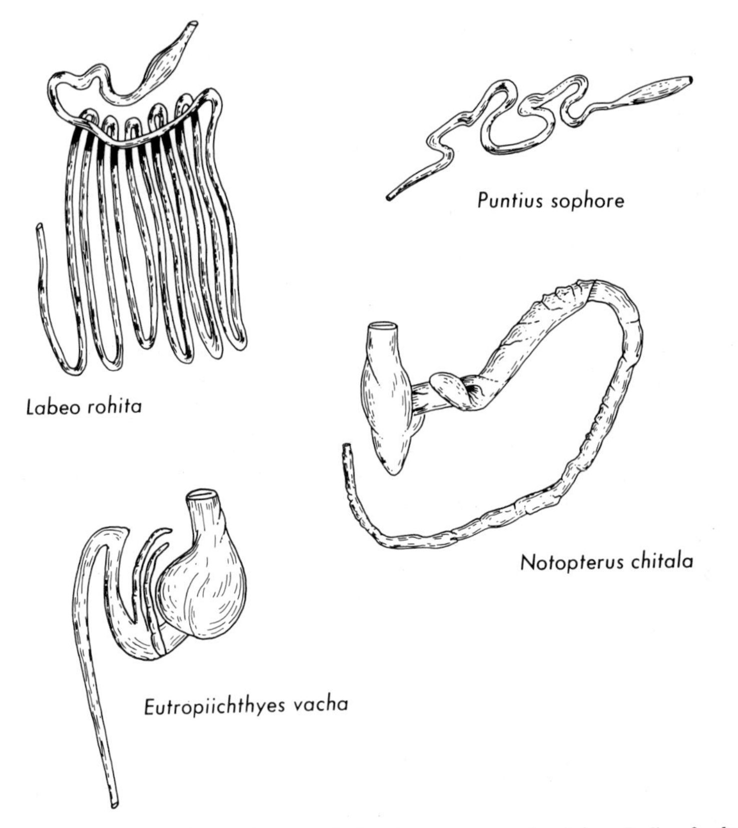

Fig. 5.1. Gross external morphology of the alimentary canals of four Indian freshwater fishes from the Uttar Pradesh. *Labeo rohita*—herbivore eating algae and soft aquatic plants; *Puntius sophore*—omnivorous, but with plants predominating in diet; *Eutropiichthyes vacha*—omnivorous, but with marked tendency to feed on larger invertebrates; *Notopterus chitala*—highly carnivorous, feeding mainly on larger crustaceans, insects, fish. (Adapted from Das and Moitra, 1956b.)

tions in other fish faunas and these associations of morphology and function clearly hold for fish everywhere, in both temperate and tropical conditions and marine and inland waters. It may be noted that fish of all sizes and species have particular upper size limits for the food items they can ingest simply because of the limited gape of their own jaws.

Fryer (1959) described several adaptations of size and form of mouth, and shape and arrangement of teeth, relative to the type of food of *Haplochromis* spp. of Lake Nyasa. *H. erichilis* eats insect larvae and nymphs from rock surfaces and to collect them it places large labial lobes against the rock, evidently using them as tactile organs to detect the movements of prey. *H. fenestratus* is a generalized feeder taking algae, microscopic animals, insects and crustaceans. Its small mouth is armed with three or four rows of little teeth. *H. polydon* and *H. pardalis* are two strongly predatory species living mainly on other fish and insects. They have similar mouth structures but the more specialized of the two is *H. polydon* which, apart from a slender body and prominent eyes, has a mouth that can open extremely wide, with sharp backward-directed teeth set in the jaws and a toothed pharyngeal bone. These haplochromids furnish an excellent example of species radiation because, though similar in general body form, they show marked differences in mouth and jaws correlated with corresponding differences in food habits. These forms are evidently the result of a recent and rapid evolution. An even more remarkable and far more specific adaptive restriction is that of *Genochronius mento* whose mouth structure enables it to eat scales and fin rays of other living fish (Fryer, 1959).

In sum then, a fish's functional morphology can often indicate the kinds of foods it eats and, perhaps just as importantly, the kinds it is unable to eat. As always, specialized adaptive structures and functions operate both to the advantage and disadvantage of their possessors, making it relatively easy for an animal to accommodate itself to some environmental feature or take ready advantage of some resource, but also restricting the animal's habits in a way that, in both evolutionary and ecological senses, may lead to blind alleys of extreme and inflexible specialization (see also Chapter 2 of Nikolskii, 1969).

However, this is not the whole story. Food habits may change as fish grow, accompanied by marked changes in the morphology of the alimentary system in early life (Chapter 2, Nikolskii, 1969). In most fish the alimentary canal, after hatching or birth, is but a simple tube. After yolk absorption, when feeding begins, the tube rather rapidly metamorphoses (as Fig. 5.2 illustrates for certain predatory fish) so that after only a few weeks the tiny fish have essentially the form of alimentary canal they retain throughout life. The major change occurs during a period in which the larval fish feed on plankton and other minute food items. Their food is still small by the time they have acquired the "adult" type of alimentary canal, but unless plankton is especially plentiful they will already be consuming foods of a new size order—such as amphipods and chironomids.

Then there are the less specialized feeders among fish—somewhat inappropriately referred to above as "omnivores'—which consume a

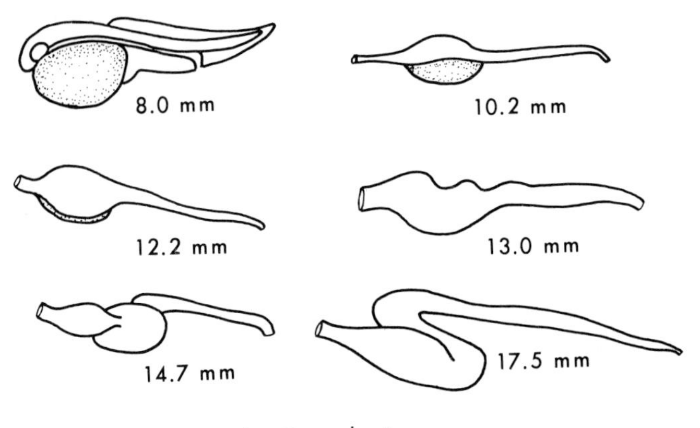

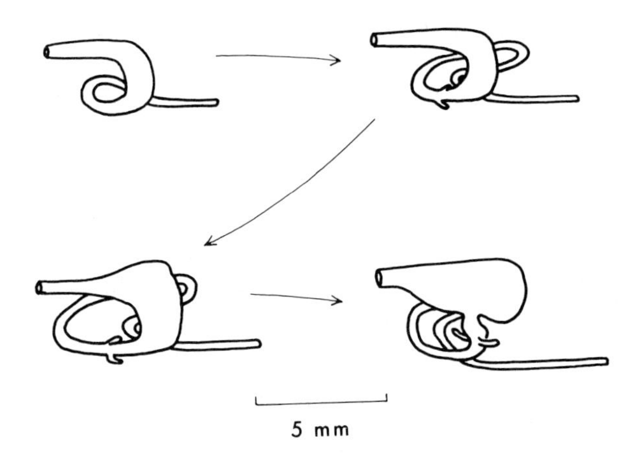

Fig. 5.2. Changes in gross external morphology of alimentary canal in growing fish. A.—Development of the alimentary canal in pike fry; shows the six main changes in shape, from that of newly hatched fry (8 mm long, yolk sac attached) to that characteristic of adult fish at 17·5 mm. (After Frost, 1954.) B.—Changes in the alimentary canal of young perch. The sequence of changes occurs in the fry stage and occupies about one month. (After Smyly, 1952.)

broad range of food materials, from plants, phytoplankton and detritus to almost every sort of available invertebrate of a size big enough to be useful. Many of them will also take fish of other species if the latter are small enough. At certain times and places, such wide ranging opportunistic feeders can seem deceptively narrow in their choice of foods, but this may be merely an indication of the ease with which the predominant food item is being obtained. As Allen (1941) pointed out "... when a feeding fish is beginning to select one particular type, the animal chosen will be one which it sees at least fairly frequently."

As a result of his conclusion that most of the generalized feeders in fish communities are facultative, Hartley (1948) suggested that a tendency to change the proportions of dietary constituents readily is perhaps a major method by which fish avoid direct competitive clashes for food, but there is little definitive evidence for this. Moreover the variable diets eaten by some fish can complicate the determination of the full feeding pattern of a species over its entire size range. To be sure of obtaining precise information about feeding habits there is usually no avoiding extended surveys of diet, feeding behaviour and the actual availability of food items.

II. COMPETITION AND GROWTH

Growth flexibility, a major fact in the life of fish, is manifested in numerous ways, some of which are certainly related to factors of space (i.e. population density) and food supply. Put broadly, fish tend to grow rapidly when food and/or space are plentiful and slowly when either or both are scarce. It is therefore often inferred that the degree to which fish are "crowded together" will affect growth. We will suppose that this influence between fish, which is manifested differently in the presence of different conditions of food and crowding, is a form of competition. For present purposes competition is defined as the state existing between animals securing supplies of the same resource from one region of an environment, resulting in an interaction that produces some actually or incipiently deleterious effect(s) on one or each of the animals (Weatherley, 1966). However, it must be emphasized again that it is only in very approximate terms that intensity of competition is related to mere population density. The total metabolic requirement of a group, cohort or population of fish determines its food need and only when that need is considered against a background of available food, does competition—which always carries overtones of some sort of struggle between individuals for an environmental resource—become meaningful.

Weatherley (1966) suggested that:

$$Ci = \frac{Me}{M_p}$$

where Ci is competitive intensity, Me is rate of utilization of food or energy by the population, M_p is rate of production of (available) food in the environment. This formula resembles that employed by Shorygin (1946) to account for competition between populations. The steps in estimating total food (or energy) requirements of populations of fish composed of individuals of various sizes have been indicated in Chapter 4. As mentioned in the last section, the estimation of M_p may also be technically exacting.

III. EVIDENCE OF COMPETITION FOR FOOD

What kind of evidence permits the inference that the intensity of competition for food increases as fish grow larger, food scarcer or space less?

The work of H. S. Swingle has long been concerned with attempts to improve fishing in American farm ponds. Swingle (1949) has described how the number of farm ponds in the U.S.A. grew from 7885 (40 000 ac) before 1935 to 1 111 253 (529 500 ac) in 1949 (note reduction in mean area). At first, Swingle favoured the bluegill bream *Lepomis macrochirus* as a pond fish, though preliminary experiments suggested that the high reproductive rate would be a disadvantage. A result, from the stocking of a 0·5 ac fertilized (chemically enriched) pond with 750 bluegill fingerlings, follows (Swingle and Smith, 1940):

Date	Average weight (g) of stocked bluegills
Mar. 25 (when stocked)	5·8
June 15	70·0
July 13	68·9
Aug. 13	56·0
Nov. 30	54·2

The bluegills grew rapidly at first so that by June 15 their average weight was 12 times greater than when stocked as fingerlings. But, because bluegills can reproduce in the summer when they reach about 35 g weight, by June they had spawned and hordes of young bluegills entered the population (2000 per pair of spawning fish) to compete with the spawners for food. By July 13 the fish originally released, weighed no more than they had a month before and by August and November averaged about 20% less than their June weight. This original demonstration of the effect of reducing space per fish (increasing population density while keeping food supply high by consistent chemical fertilization of the water all summer) and thus raising the intensity of competition for food might, of course, be interpreted in other ways. However, those bluegills weighing less than 1 g consumed the same food as the larger bluegills and, even in very densely populated ponds, large bluegills refused to eat their own young. It was therefore generalized that, even if many years are available, populations of bluegills cannot regulate their own numbers effectively in ponds. Once the fish originally released have reproduced, direct competition between age groups and lack of cannibalism combine to keep population density high and growth rate low, even when food production is high. This experiment was repeated by Swingle many times in various forms with similar results. Further illustrations were published of the relationships between density, competition, food and growth, which offered two sorts of evidence; the first of which dealt with different food supplies (Swingle and Smith, 1942). A pond was stocked with 1500 bluegill fingerlings per acre which after one year averaged 4·0 oz in fertilized ponds in which food was very plentiful, but only 1·1 oz in unfertilized ponds in which food was relatively scarce. The second sort of evidence was of the effects of different population densities; the first example of which was of 1500 bluegill fingerlings per acre, which at one year averaged 4 oz each, compared with an average of 0·02 oz each in a population with an initial density of 180 000 fingerlings per acre. The second example of the results of changing density was based on 6500 bluegills per acre in three fertilized ponds. After 2½ years the populations in two of the ponds were reduced to 3200 and 1300 per acre. After another six months the mean weights of bluegills in the three ponds were (highest to lowest population density) 0·9, 1·8 and 3·7 oz respectively. The third example consisted of stocking bluegills at 6500 per acre in fertilized ponds in which the fish averaged 0·8 oz at six months. However, as a result of subsequent spawning, the mean size of these stocked fish two years later was still only 0·9 oz. This is essentially another example of the first observation by Swingle and Smith referred to above.

From further publications (Swingle, 1951) it appears that, in unexploited populations of bluegills in ponds, of those stocked as fingerlings about 20% die per year, largely irrespective of initial stocking rate (population density).

It is interesting to compare these results to the well-known experiments of Nicholson with monospecific dipteran cultures (Nicholson, 1954, 1957), in which there was a regular cyclic fluctuation in numbers in established cultures fed regularly and either left alone or killed at planned rates. Many population theorists have accepted Nicholson's explanation of his results and it has been widely regarded as applicable in principle to various animal populations. However, none of the features Nicholson found are seen in Swingle's work in which population stability seems virtually final in remaining nearly unchanged for many years once certain events have occurred. Nicholson's explanation of the notably regular density oscillations among his experimental dipteran populations is as follows (Nicholson, 1954, pp. 20-21):

> It will be observed that significant egg generation occurred only when the adult population was very low. At higher densities competition amongst the adults for the ground liver was so severe that few or no individuals secured sufficient to enable them to develop eggs. Normal mortality, therefore, caused the population to dwindle until the consequent reduced severity of competition permitted some individuals to secure adequate liver and so to lay eggs. As it takes more than 2 weeks for the eggs so generated to give rise to new adults, the population continued to dwindle for this period, during which many more eggs were generated, for competition among the adults for ground liver continued to slacken. The eggs then generated in due time gave rise to new adults, which led to a rapid increase in the adult population, and the resultant overcrowding caused virtual cessation of egg production. A new cycle of oscillation then began.

In contrast to such conditions, the fish in Swingle's ponds have large juvenile populations of fixed size (i.e. unable to grow) and small adult populations (also unable to grow) and no reproduction at all. It is probable that over several years enough smaller fish would die to enable at least some mature females to reproduce again, if any of the original females was surviving. Otherwise, as the population gradually decreased through mortality some of the smaller fish would eventually reach spawning size themselves. However, we could suppose that young fish thereby added to the population, though they would increase the number of age groups, would also serve to maintain the competitive pressures which arrest growth.

Cessations of growth can also be found in fish populations in nature, though these generally take a less severe form than those in

Swingle's ponds. However, Swingle himself was attracted by the frequency with which "stunted" individuals of certain species are found in wild populations. Nicholls (personal communication) detected arrest of growth in trout which had reached a certain age in certain Tasmanian lakes. Deelder (1951) described a cessation of growth among perch in irrigation channels in Holland. Both these examples may be due to competition for food among certain size groups. Swingle's pond populations were, however, characterized by general severity of competition—no fish were growing much, whether old or young, large or small. In a sense this was an outcome of the small size and simple conditions of these ponds and, of course, stems directly from the fact that all the fish were consuming the same food. In larger, more complex environments, fish of different sizes would be more likely to manifest major or minor preferences for different foods and gratify these preferences through differences in distribution (Nikolskii, 1969). So, though Swingle's demonstrations are illuminating, they should be applied cautiously to wild populations. Possibly similar reservations should apply to Nicholson's findings with respect to cultured fly populations (Andrewartha and Birch, 1954; Andrewartha, 1961).

Other evidence can be adduced as to the relation between food and space in governing growth through the influence of intensity of competition for food among fish. Hile (1936) reported an inverse relationship between growth and population density of the cisco *Leucichthys artedi* in four highland lakes in Wisconsin. Hile ruled out as a major factor the fact that the growing season was a little longer in the lakes where the best growth occurred, though this certainly introduced a confounding effect. It also appeared that the poorest growth was in the most eutrophic of the four lakes. Hile usefully reviewed other evidence on the effects of density on fish growth in this important early study. Beckman (1940, 1942, 1946, 1948) reported marked changes in the growth rate of several species of fish after reductions in lake populations through poisoning and the mass mortalities of severe winters Rose and Moen (1952) described an increase in growth rate of fish of sporting value following large-scale selective elimination from a lake of fish without recreational value. Both "sport" fish and "coarse" fish were consuming many of the same sorts of food organisms and thus presumably competing for food.

Le Cren (1958) found an increase in growth rate of perch in Windermere during the course of a trapping programme extending over many years. However, this effect apparently had a complex origin only partly attributable to lowered intraspecific competition for food (see Chapter 4). Figure 5.3 shows the changes in population density and biomass from 1940 to 1955. These values are actually *ratings* of population density and biomass derived from catches by standard perch traps though there certainly must have been a vast decline in absolute numbers and biomass of perch in Windermere.

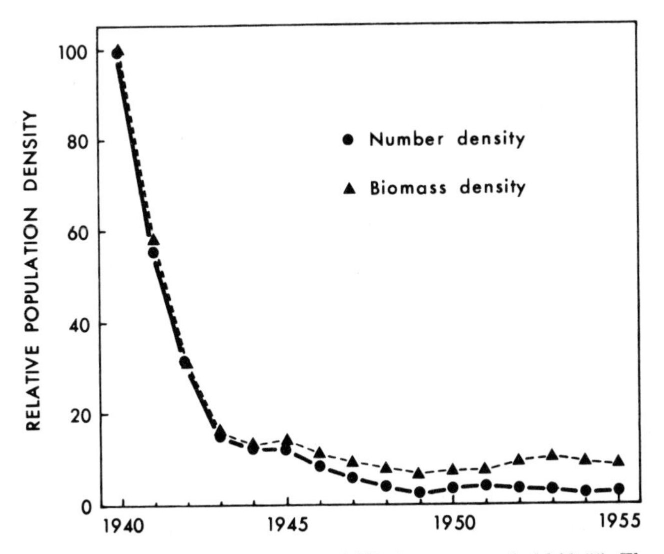

Fig. 5.3. Changes in population density of Windermere perch 1940-55. The geometric means of the catches per trap made in successive pairs of years from 1940 + 1941 to 1955 + 1956 expressed as a percentage of the 1940-41 value. The mean for each pair is plotted against the first year, as indicating the population density in the summer growing season immediately following the trapping season. (After Le Cren, 1958.)

Over the experimental period mean size of trapped and angled perch showed a convincing reciprocal relationship with biomass. While biomass was being reduced by 95%, mean weight of trapped males and females rose approximately by 200 and 4-500%, respectively. Figure 5.4 expresses this as three curves depicting perch growth in 1941, just after trapping began, and again in 1944 and 1949. The most important change in growth did not occur until perch were three years old, when a significant increase appeared, maintained throughout subsequent life. The increase was exponential and adult growth rate was highly correlated with time from the start of the fishery, but not with changes in population density after 1941. (Remember that Le Cren showed by statistical analysis that much of

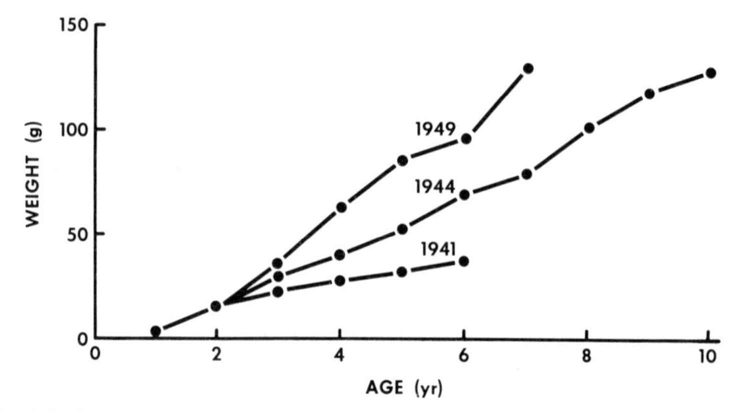

Fig. 5.4. Weight-for-age for male perch in Windermere of the combined year-classes before 1941 and the 1944 and 1949 year-classes compared. (Adapted from Le Cren, 1958.)

the yearly variation in growth was attributable to temperature (Chapter 4)).

Le Cren believed there was plenty of planktonic food for young perch, which may have explained why changes in population density apparently did not affect growth of that life stage. It seems more certain that the benthic food for small perch more than 10 cm long, when they generally cease to feed on plankton, did increase as a result of the reduction of the perch population. This would help to explain the major growth stimulus in the third and fourth years of life. Yet, after 1942 and 1943, growth of fish already sexually mature did not rise further, despite the sustained reduction of numbers and biomass. Le Cren thought that "... perhaps these further stimuli were not adequate to induce further increases in a growth-rate normally fixed and stable by this age."

It is obvious that in any large natural water body, changes in growth rate of a particular species will be affected by populations of other fish, which in Windermere include minnow, stickleback, eel, pike, trout and char. Trout in particular, feed on many of the same animals as perch (Frost 1946). Le Cren (1958) estimated the trout biomass before 1941 as no more than 1% that of perch, but between 1944 and 1955 it approached that of the greatly reduced perch population, while remaining less. That is perch were perhaps 20 times less abundant (biomass) and trout nearly five times more abundant than before 1941. The operational problems of studying so large a population are obvious, but it does seem slightly regrettable that it was not possible to make some assessments of plankton, benthos and populations of other fish. Much that remains conjectural might thereby have been clarified. However, the experiment was in many ways notable, almost unique in its duration.

El-Zarka (1959) reported a markedly slower growth rate in yellow perch *P. flavescens* in Saginaw Bay, Lake Huron in 1943-45 as compared with 1929-30. He thought it was probably attributable to the estimated seven-fold increase in population density that had occurred over the same period. Abundance indices of four out of five other major fish species in Saginaw Bay also showed more or less pronounced increases and, since interspecific competition may have been involved, the decline in perch growth may have had several causes. El-Zarka was unable to implicate physical factors such as turbidity, temperature or precipitation as conclusively affecting growth.

Other, less direct, demonstrations of the probable effects of manipulating population density have been given by Alm (1946) and the well-known experiments on commercially-important marine fish (see Beverton and Holt, 1957, for comprehensive, critical review). In some of these studies (e.g. Alm's) growth improved as a result of transference of fish from environments in which they grew slowly to new environments. Though this again clearly exemplifies the flexibility of fish growth and certainly indicates that the new environment was in some way more "favourable" it is difficult to decide whether food, population, new physico-chemical conditions or a combination of all these promoted the better growth. It is very hard to be sure of the superiority of a feeding ground by mere examination, yet that is what many studies of benthic food have amounted to, in comparison with recent efforts to estimate primary, secondary and tertiary production objectively. Similarly, the demand of a fish population for food cannot be computed merely from estimates of population density. Quantitative delineation of population structure is needed, plus some knowledge of whereabouts of all size groups of the population in space and time. Only then can a reliable picture of the food demands of the population be formed and an impression gained of how adequately these are being met by the food available (see again Chapter 4). The more useful and reliable experiments are those in which growth rate is estimated before and after a change in population density or biomass or after a measurable change in food supply in the same water body.

Swingle and Smith (1939) showed quite clearly that, by increasing food supply while holding space constant, a marked increase in growth per fish could be expected. Swingle made use of this in recommending stocking rates for farm fish ponds that varied according to whether the pond was fertilized (highly productive of food organisms) or unfertilized (poorly or moderately productive). In the productive ponds about four times as many fish were able to grow at about the same rate as in the unproductive ponds (see Swingle and Smith, 1942, 1950).

A surprising number of attempts have been made to increase production in lakes and even in semi-enclosed sea lochs by chemical enrichment (Chapter 9). The approach has usually been to supply inorganic nutrients for phytoplankton, thus removing a common limiting factor to production and promoting a phytoplankton "bloom". Improved growth of fish has been expected to occur through augmented food chains. Most of the attempts have considerably affected the biotas of the water bodies and improvements in fish growth have sometimes been demonstrated (Chapter 9).

Before concluding this section the following observation may be noted: "The importance of population density in influencing growth rate of fish has long been recognized . . . but the exact relationships between numbers and growth and the role of the food organisms are obviously complex and still poorly understood" (Le Cren, 1958).

Such poor understanding will persist until it is more generally realized that it is not mere numbers (i.e. population density) and, therefore, not space per individual nor the richness of the food supply itself that influences growth, but rather intensity of competition for the food.

Backiel and Le Cren (1967) reviewed density relationships in fish, basing their evaluation mainly on evidence from *Cyprinus carpio* in ponds, brown trout in a screened section of a stream (after Le Cren, 1965) and on growth of haddock in the North Sea in populations of different densities. They proposed that the general form of the relationships between growth rate and population density was:

$$\frac{dG}{dN} \text{ or } \frac{d(\Delta W)}{dN} \propto \frac{1}{N}$$

This implies that changes in population density at low levels of population are associated with much greater changes in average growth rate of individuals than when corresponding changes in density occur at high levels of population. Backiel and Le Cren believe that this formula describes these relationships in a realistic way. However, in fish, in which growth in the individual is plastic, it is unlikely that any attempt to analyse growth effects in terms of density alone can lead to much insight into the nature of *this* problem or to a realistic description of the mechanisms of competition-dependent growth variability in fish, which must take into account the size and activity of the fish that are competitively interacting. Unless we are dealing almost entirely with a single age group in which size varies over only a narrow range it will be best to consider *density* in fish populations as not being generally and systematically related to growth.

In Chapters 2 and 4 growth was examined against the background of the conflicting requirements of standard metabolism, excretion, active metabolism etc. Weatherley (1966) gave a simple model of this problem which may be useful in ecological population problems. This is reproduced here (Fig. 5.5), with the warning that the parameters suggested are both arbitrary and schematic. Procedures for measuring the values of the various fractions of metabolism were discussed in Chapter 4 and it has also been pointed out (and see Weatherley, 1966) that in an actual population, data on specific mortality and recruitment rates, immigration and emigration etc. would be required (Chapter 8). We would also need metabolic data, not on individuals representing the average size in the population, but representing all the different size groups, together with estimates of the numbers of individuals in those size groups. If social factors such as inhibitory or stimulatory effects of population density itself or factors associated with schooling, are suspected as influencing any aspect of metabolism—e.g. activity—they should be tested in adequately controlled experiments. All this implies a formidable programme, which is, however, the only proper way of clarifying the effects and importance of competition relative to the manner and velocity with which matter and energy are utilized in the processes of growth.

Some of the requirements for suitable laboratory experiments on growth were followed by Magnuson (1962) in his analysis of competition and growth in juvenile siblings of the cyprinodont *Oryzias latipes*—the medaka. Magnuson found that, "Growth depensation is no greater in populations at densities up to 16 fish per liter than . . . raised in isolation under the same conditions (food in excess, accumulation of water-borne growth inhibitors prevented)." Magnuson found aggressive behaviour and size hierarchy effects were significant not when food was in excess but when it was limited. In the social hierachy which then develops larger medaka dominate and grow faster than subordinates, although the "dominant has no advantage if no food or excess food is supplied." Magnuson (1962) also showed that aggressiveness was conditioned by hunger and the behavioural attitude of medaka towards food depended on whether

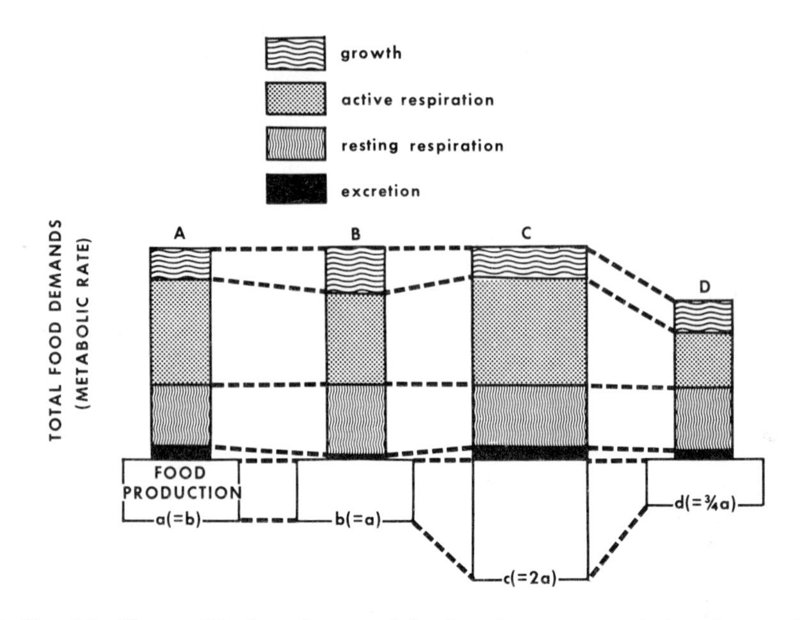

Fig. 5.5. The partitioning of ingested food or its energy equivalent in two fish populations, A and B. The food supply is the same to each. It is proposed that higher rates of activity (as in A) lead to reduced growth, reduced contributions of resting metabolism and increased contributions of excretion to total metabolism. Yet the ratio $Me/M_p = Ci$ remains the same for both populations because total metabolism is the same.

In population C the food is supplied at twice the rate as in A and B, so C may show the same proportional partitioning of food between growth, active and resting respiration and excretion as A or B but twice as much of each of these components of metabolism. Again, Ci is the same as in A and B.

Population D has the same total quantity of food going into growth as has A, though the actual food supply is only three-quarters of what is available to A. This growth equality is possible if it is supposed that the demands of D for activity are considerably less than those of A. Despite this difference in food supply combined with the same "demand" for food by growth, again, intensity of competition for food within population D ($Me/M_p = Ci$) will not—in the example given—be higher than in population A.

These several examples show that ecological appraisals of the significance of growth and competition are not to be inferred with certainty from, for example, measurements of growth alone. They are intended as a model of the complexity of interaction between different components of metabolism. (Modified after Weatherley, 1966.)

the food was in localized masses or evenly distributed and on the complexity of the environment: "If there is one semi-isolated subsection in the environment for each fish and if limited food is evenly distributed, both the dominant and the subordinate grow equally well." Many other points investigated by Magnuson clarify the behavioural complexities of "competition". He sees aggression as a force which helps to distribute fish evenly throughout their environment when food is dispersed, but not so strong as to prevent them from coming together to feed if food is clumped. If this very interesting study were extended to older age groups and larger species of fish many important insights into the nature of competitive interaction would accrue.

Experiments of Chen (1965) and Chen and Prowse (1966) suggest that the absolute amount of space available to a fish may influence its growth rate independently of the relative amount of space available. At the Tropical Fish Culture Research Institute (Malacca) they released into ponds all male hybrid *Tilapia mossambica* at the rate of 1000 per ac. All ponds were made as similar as possible except for their areas which were 1·0, 0·5, 0·25, 0·1 and 0·01 ac. Chen and Prowse reported a notably superior average growth rate for fish in the largest pond compared to all others with a slightly, but distinctly greater growth rate in the 0·5 ac pond than in the 0·25 ac pond. Observations on the other ponds were discontinued after six months and are therefore inadequate for comparison. If, as they suggest, Chen and Prowse have demonstrated that the actual absolute size of the water body fish inhabit—quite apart from population density—influences growth, this could be an important factor in evaluating the relations between competition and growth. Though this work is very interesting it has the disadvantage that only one pond of each area was used; moreover, the assumption that each was of identical productivity per unit area or volume appears questionable. If the observed result is of general applicability, however it is of considerable importance. It would be highly desirable to repeat and extend the experiment to more ponds and species and with a more searching attempt made to ensure that food production is truly comparable in ponds of all sizes. There appears to be no very plausible explanation of the observed growth differences at the moment.

IV. THE MEANING AND SIGNIFICANCE OF COMPETITION

Many modern textbooks of ecology give a fairly full critical consideration of the notion of competition, in recognition of the assumed importance in ecological situations of competitive interactions. However, "competition" has also become a very contentious term so that clarification of its use in this instance is necessary.

Milne (1961) summed up the sometimes genuinely analytical but frequently semantic or polemical debates that the problem of competition has generated. He also furnished his own definition, allegedly shorn of most epistomological shortcomings, but not quite satisfactory for present purposes, "Competition is the endeavour of two (or more) animals to gain the same particular thing, or to gain the measure each wants from the supply of a thing when that supply is not sufficient for both (or all)."

Weatherley (1966), while considering Milne's definition superior to most others, proposed the following modification: "Competition is the state existing between animals securing supplies of the same resource from one region of an environment resulting in an interaction that produces some actually or incipiently deleterious effect(s) on one or each of the animals."

The subject of competition is still a live ecological issue as is indicated by the recent searching essay of Miller (1967), who noted that dictionary definitions of competition are inadequate to express ecological meanings and that many ecologists have pondered the appropriate application of the term. He adopted a version of the early definition of Clements and Shelford (1939): "Biological competition is the active demand by two or more individuals of the same species population (intraspecies competition) or members of two or more species at the same trophic level (interspecies competition) for a common source or requirement that is actually or potentially limiting."

Against the arguments and attempted definitions of the ecologists who at least hold in common that competition is a very important process within and between populations are arrayed the counter-arguments of Andrewartha and Birch (1954) and Birch (1957). Basing their standpoint to some extent on the writings of Thompson, W. R. (1939) and Ullyett (1950), Andrewartha and Birch tend to oppose the use of the word in any generalized or non-specific sense, in case such a use may deflect ecologists from attempting to analyse the real effects of interactions between animals. Thus, where a group of animals is too numerous for its food supply, some or all might be seriously affected and even starve. They would contend that it is an insufficiency of food for the number or bulk of animals, not competition, which leads to death. We must approve their insistence on rigorous analysis of the mechanics of animal interactions and their adherence to a simple, consistent terminology for ecological processes, so that analysis will not become needlessly confused by semantics. Nevertheless, it appears safe and convenient to use a term such as competition if its meaning is stated precisely as in the definitions cited above.

It certainly seems unsatisfactory to define competition without giving an indication of its results contrary to the strictures of Milne (1961), who believes that, while competing is something animals do against each other, a definition of it should not be enlarged to suggest possible *effects* of their struggle. But if we accept this we shall be drawn to such logical absurdities as the "co-operative competition" Allee *et al.* (1949) described (see also Birch, 1957).

Andrewartha and Birch (1954) and Birch (1957) rightly criticized some very generalized but superficial uses of the term competition, which have suggested that instances of predation, parasitism and other effects of one species upon another may all be similar because all may be accounted for by the same mathematical formulas. They have also severely criticized the position of Lack (1954, 1966), who suggested that the fact that birds avoid occupying identical "niches" (actually mainly nest sites, foods and feeding areas) was an evolutionary expression of the Gaussian axiom that two or more species cannot continue to occupy the one "niche" because of a principle of competitive exclusion. Lack's position is that long ago different species of birds may have temporarily occupied similar space, but that selective influences worked against their continuing to do so. The hypothesis is attacked on the grounds of insufficient evidence that the birds in question ever did occupy exactly similar "niches" or that competition for requisites forced their distributions apart in the course of evolution.

Although Andrewartha and Birch (1954) were critical of ideas (such as Nicholson's, 1933, 1954) that animal numbers in nature are generally regulated through density dependent factors, they do not deny the occurrence of competition. Indeed their definition is as follows:

> We shall ... say that competition occurs whenever a valuable or necessary resource is sought together by a number of animals (of the same kind or different kinds) when that resource is in short supply; or if the resource is not in short supply, competition occurs when the animals seeking that resource nevertheless harm one another in the process.

The definition given by Weatherley (1966) is close to this. However, it derives from an earlier paper on the subject (Weatherley, 1963) in which it was assumed that competition does not begin at some critical level of population density, but may be incipient in an activity such as consumption of a food source common to two or more animals. This may lead, in fish, to a reduction in growth rate not necessarily an immediate disadvantage, but likely to become a deleterious effect in the face of a severe or maintained competition for food.

A single animal can be short of food to an extent that some vital process becomes limited but we can only think of *competition* as limiting food when an animal is reducing the food supply of another—and probably having its own food supply reduced in turn. An individual may suffer a shortage, but the shortage is linked to competition only if it has been brought about or exacerbated by other individuals. It must not, however, always be assumed that one or more animals consuming the same sort of food in one environment are competing. They may be too far apart to do so. The definition (Weatherley, 1966) specifically refers to animals securing supplies of the same resource from one region of an environment. Competition may frequently be occurring in ways not at all obvious. For instance, fish of different species or populations may feed at the same place on the same foods but at different times, so that they never directly contact each other. Weatherley (1966) pointed out that the nature of competitive interaction varies. Sometimes it may be in the form of a direct struggle between animals for some resource. In the case of food it probably more frequently assumes the form of a behavioural inhibition restricting feeding territory or activity (Magnuson, 1962). For an ecologist, however, the precise character of the behaviour is less important than the nature and severity of its effects.

In sum, then, when applying $Ci = Me/M_p$, care must be taken that the fish are in a situation (part of a population or community) in which it is realistic to expect competition to occur. In addition, one needs to be sure that the fish really are eating the same food. Lastly, there must be *at least* two fish involved.

Is there, finally, some peculiarly obscure difficulty associated with the use of the term competition that will continue to stand in the way of its more general use by ecologists as a fairly precise definition of a particular kind of process? Certainly, it is very difficult to know what is meant by the term in many contexts. Some biologists appear to think of it as descriptive of comprehensive interactions between organisms that take place not only in relation to particular resources but which encompass all the commonly occupied space-time plus resources-in-general of their common ecosystem. In a sense Wynne-Edwards' (1962) treatise—which attempts to demonstrate how animals *avoid* the direr forms of competition— may be a kind of massive reaction against this too-sweeping concept of competition.

Perhaps, if the users of the term competition would try always to specify precisely what they mean by it and the sorts of context in which they want to apply it, many of the doubts relating to its use would be removed. It might even become apparent, to those who went to this trouble, that they were all talking about essentially the same sort of process!

V. THE CONCEPT OF THE NICHE IN RELATION TO COMPETITION AND GROWTH IN FISH

The idea of the niche (Elton, 1927) and the consequences of the findings of Gause (1934) and Volterra (1931) have both complicated and refined the problem of competition. Generalized competition, a kind of matching of the various demands of all stages of an organism's life cycle against those of other organisms, appear to be unfeasible and yet many ecologists can apparently envisage various species occupying similar territories and needing similar materials in space and time; hence the well-known idea of the niche as an "*n*-dimensional hyperspace" (Hutchinson, 1957, 1965) in which the environment of an organism is conceived in terms of values for a range of variables each plotted on an axis joining the axes of other variables. "Along each axis any particular species has a tolerance range in which it can live successfully. By defining the tolerance range of a species along *n* environmental axes, a volume is generated in *n* dimensions that represents the potential survival area of the species" (Wuenscher, 1969). In such a use of the niche concept, however, we would be adopting a view which required competition to be defined in terms of similar demand for all kinds of requisites at similar times and places. As Bodenheimer (1958) warned, if the concept of the niche is framed too broadly and inclusively its ecological usefulness is nullified: "It is erroneous to expand it to a meaning synonymous with ecoworld . . .", which would be virtually to make it a substitute for the concept "ecology" itself. Weatherley (1963) proposed that the niche be regarded merely as the nutritional role of the animal in its ecosystem, that is, its relations to all the foods available to it. This definition (drawing heavily on the ideas of Elton, 1927) is narrow, but allows the ecologist to compare the nutritional role of one species in its ecosystem with other different—but nutritionally comparable—species in other ecosystems; see also De Bach (1966) and Miller (1967).

A majority of ecologists feel that between two or more species competition will be at its most intense where both or all are occupying much the same space and consuming the same resources (e.g. food) concurrently. Indeed this idea dates back to Darwin (and see Andrewartha and Birch, 1954; Birch 1957). The notion has an intuitive appeal, but let us now consider whether the facts about fish make it likely that it can really be applied to them.

We have been referring partly to intraspecific and partly to interspecific competition, but there seems no reason to make this separation so far as fish are concerned. Let the discussion be limited to the struggle for a single resource: food. Many studies of the foods consumed by more than one species of fish within a single ecosystem have emphasized both interspecific similarities and differences. Great differences are normally taken to indicate that competition for food is negligible and marked similarities as indicating actual or potential interspecific competition. Fryer (1959) referred to such a situation among herbivorous fish in Lake Nyasa "in which a number of closely allied species live together in the same microhabitat and eat the same food." This situation illustrates the dilemma of the ecologist who accepts the idea that, where food or space are limiting, two or more species cannot continue to occupy the same "niche". Fryer stated: "Such co-existence is at variance with the so-called Gaussian hypothesis and seems possible because of a superabundance of the algal food on which the species feed." Fryer's interpretation of his observations may not be unreasonable but it rather ignores that the essential consequence of competition for food among fish—in either its intraspecific or interspecific forms—is reduced growth rate of individuals rather than density-dependent mortality. If two or more species of fish do occupy the same microhabitat (space) and eat the same food which is *not* superabundant it is not clear why eventual extinction of all except one species should be expected. If the spawning of the females of one or more species were impeded through overcrowding of breeding grounds that might, admittedly, reduce recruitment. But there is no obvious reason why this would lead automatically to extinction of such species.

A summing-up of this question of the niche, then, proceeds along the following lines. If the Gaussian principle of competitive exclusion of species occupying the same microhabitat and sharing a food supply be taken to have the important general application some ecologists have suggested, there can be several ways in which fish may avoid its consequences. In the first place, for multispecific fish communities of no fixed feeding habits (Hartley, 1948; Maitland, 1965; Mann and Orr, 1969), competition could probably never be more than a fleeting problem. Most of the fish of such communities readily change their diets and grow satisfactorily on such a wide range of foods that they are unlikely to suffer a prolonged disadvantage from food shortage induced by competition for a particular food. Moreover, in these communities, differences in behaviour, activity patterns and habitat preference make it unlikely that two or more species persistently clash competitively. Secondly, it may be possible that at certain places and times, Fryer's (1959) argument will hold: a potential competitive encounter between species will be avoided because the food—although the same for all species—is so abundant that all have excess. Thirdly, even if interspecific competition for food is unavoidable and extensive, growth flexibility may enable fish to avoid its more severe effects. Probably in the majority of cases competition between fish will be discontinuous, restricted to more or less acute convergence of requirements for a resource for limited periods.

Lastly, let us briefly turn to the study by Keast and Webb (1966) of mouth and body form of fish in a small lake and their feeding on the organisms available to them. Their major conclusion is similar to that of Lack (1954, 1966) with regard to British birds which, as mentioned earlier, has been criticized by Andrewartha and Birch (1954) and Birch and Ehrlich (1967). Keast and Webb hold that, "In 14 cohabiting . . . species . . . mouth and body structures combine with food specializations and habitat preferences to greatly restrict interspecific competition within the fauna." They further state that ". . . the structural characteristics of food species are related to their food niches" (a claim which might, with advantage, have been reversed) and that "whereas structure does not prevent food overlap in cohabiting species, it does serve to reduce interspecific competition because most species have specializations that place them at an advantage in certain situations, or in the obtaining of certain foods."

Keast and Webb criticized Hartley's (1948) work, believing that their observations on degree of specialization of mouth structures, location and type of food and behaviour involved in obtaining it, conflict with Hartley's claim that many fish change their diet over a period. On the other hand, Keast and Webb admit that few of the fish they examined were so specialized as to prevent them from utilizing at least three different sorts of food, adding that it is advantageous for species to be able to ingest a food when it becomes temporarily superabundant. Indeed, they record that the 14 species of fish examined did often eat many of the same kinds of organisms. It therefore seems likely that some degree of competitive interaction could occur between them. Since the study did not include attempts to evaluate ecologically either the strength or absence of such interaction it has not nullified the importance of competition, though it has added to the demonstrations of ways in which its power could be ameliorated.

6 Growth and Maintenance of Populations

I. SURVIVORSHIP AND AGE/SIZE STRUCTURE

Deevey (1947) first drew general attention of ecologists to the value of life tables. His definition of the life table emphasized that: "Beginning with a cohort, real or imaginary, whose members start life together, the life table states for every interval of age the number of deaths, the survivors remaining, the rate of mortality and the expectation of further life." Few previous studies had rigorously satisfied the major requirements for construction of accurate life tables and it is perhaps interesting to note that life tables from human actuarial statistics, despite their abundant data, were not among these.

Since many assumptions about mortality patterns may be both unjustified and untrue, it is important to be able to recognize age-classes or cohorts and the development of reliable methods for this is an important part of ecological population study (see also Chapter 3). It is, for example, necessary to determine rates of migration into and out of populations with some certainty in order to establish the degree to which the characteristics of an age-size group are due to mortality. Genuine life tables can be established only after quantitative study of single age groups, preferably from birth to the death of the last individual. The best material for this is the laboratory population or, in nature, sessile animals (e.g. barnacles). However, mobile animals such as non-migratory terrestrial mammals inhabiting a clearly identifiable range, among which the major causes of death over the whole life cycle are well known and in which skeletal or other remains permit age at death to be estimated, can also furnish very satisfactory material. Deevey correctly indicated that fish populations, though much studied, are among those for which rigorously based life tables have rarely been determined, because fish are usually hard to sample regularly in a satisfactorily quantitative manner. The problem is further complicated by the fact that, unlike many organisms, fish often display "outrageous fluctuations in strength of year-classes". Nonetheless, many partial life tables and survivorship curves which are operationally extremely useful have been constructed for fish. A major difficulty with fish is their often considerable life span. While lake or pond populations can sometimes be kept under scrutiny for several years of regular sampling, year-classes often cannot be followed properly throughout their full lifetimes.

Figure 6.1, based on Deevey (1947) and Slobodkin (1962) shows four theoretical survivorship curves. Deevey attempted to find examples of types I, II and IV from actual populations of animals, but fish populations cannot be consistently represented by any of the four types over the whole life cycle. Many freshwater salmonid populations feature high early mortality, approximating to the curve IV type, followed by steady annual mortality (constant fraction

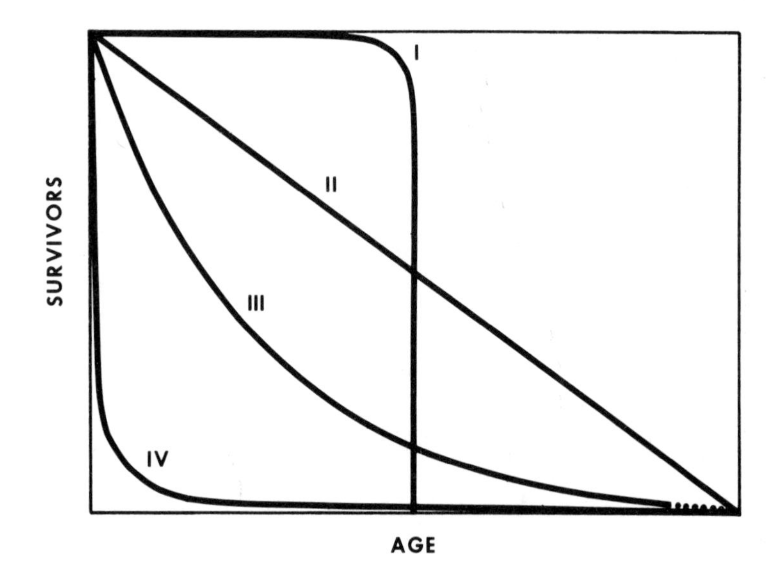

Fig. 6.1. Four hypothetical types of survivorship curve. Curve I is the "physiological" type, in which nearly all mortality occurs at one particular age (laboratory cohorts sometimes manifest this pattern of survival). Curve II depicts the effect of a constant percentage of the original number dying each year. Curve III results from a constant fraction of the cohort, remaining at the beginning of each age interval, dying in each successive age interval. Curve IV is the result of heavy early mortality with drastically reduced mortality thereafter. (After Deevey, 1947 and Slobodkin, 1962.)

dying) for a number of years (curve III), with a slight reduction in mortality rate in the last years of life (Allen, 1951, 1962; Nicholls, 1958a, b, c; Shetter and Hazzard, 1939; McFadden and Cooper, 1964). By contrast many "salmon" species die after spawning over a rather narrow size range, so that the latter part of their survivorship tends to approximate to curve I. Thus, although "standard" curves may be fitted to parts of the lives of fish, none of these can adequately account for the entire survival pattern.

Survivorship curves help in predicting the extent and effect of recruitment in enabling us to discern proportions of the population that are reproducing and, therefore, to estimate the relative reproductive contributions of various age groups. Comparisons of survivorship curves of different populations of a single species may suggest the nature of the various factors influencing mortality, which may also affect different age groups to different extents (see Slobodkin, 1962).

Much of the literature on fish populations deals with effects of changes in fishing intensity (which is, of course, a particular form of predation) on populations that were presumably in steady state originally. To say that a population is in "steady state" means that such changes in growth, biomass, recruitment and mortality as occur, do so with a regular pattern and magnitude so that at a particular point in each phase of a population "cycle", age and size structures will tend to be similar. Since many, probably the majority, of the world's larger fish populations appear to have a reproductive period of more or less regular occurrence and duration the pulse of this, also, will tend to have a reliable frequency (see, however, Nikolskii, 1969, Chapters 2 to 8). The present usage of the term "steady state" roughly approximates that of Beverton and Holt (1957), although they intended it to describe the dynamic condition of a fishery, whereas here it is extended to cover the broader concept of the population, whether or not exploited by man. The essential point of Beverton and Holt's definition, and of the present one, is that both mean primarily that the fishery (or fish population) "is not in the process of changing either in character or size."

However, this book does not specifically treat fishery problems but is concerned with fish ecology, so that this section is not so much concerned with populations in steady state as with an attempt to outline the events likely to follow a first, or colonizing, entry into a water body of a few fish—perhaps only a single breeding pair—capable of breeding there. The great practical problem of following the results of such a beginning in satisfactory detail ensures that this sketch is tentative and hypothetical. However, it appears reasonable to suggest that a sigmoidal curve of increase in numbers will describe the situation. Figure 6.2 shows such a curve for the increase of a fish population in a new, spacious environment and indicates the factors leading to rapid early growth of individual fish (not the population) and much slower subsequent growth. Biomass

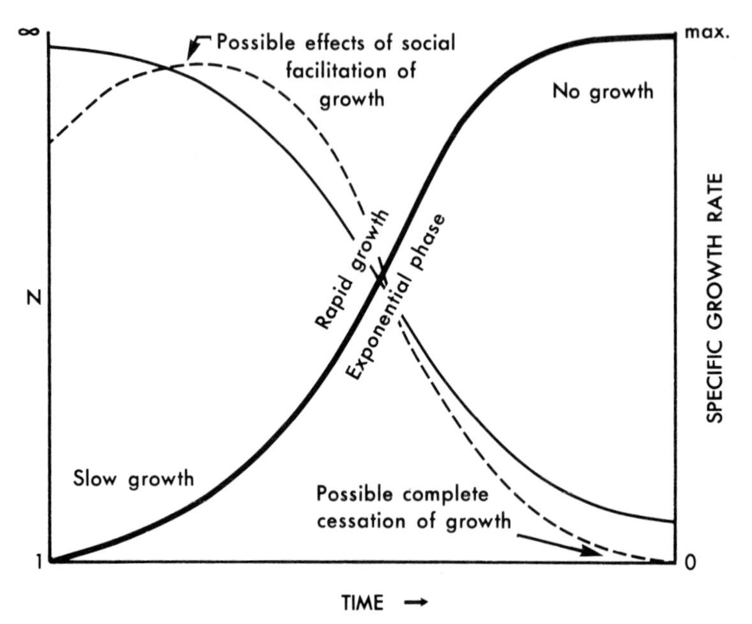

Fig. 6.2. The curve of increase (heavy line) in numbers of a fish population in a new environment in which neither food nor space are initially limiting. N = 1 indicates that the population begins with a single fertilized female. It is assumed that growth rate of individual fish will be rapid initially (light line), declining later due to rising intensity of competition for food. A third curve (broken line) indicates the possibility that growth of individuals may sometimes be maximal at population densities (or biomasses) greater than the initial (minimal) value, and also that, under certain conditions of severe overcrowding or food shortage, growth may cease completely. (Original.)

increase would, like the growth of numbers, probably also follow an approximately sigmoidal form although its precise shape would vary considerably. The mean size of fish would tend to decline as both numbers and intensity of competition build up, because of the decrease in the mean specific growth rate. Figure 6.2 allows for the possibility that growth rate may sometimes be highest when population density (or even biomass) has exceeded the initial (i.e. minimal) value, on the assumption that there may be an optimal population density for growth that, in some instances, is greater than the minimal population density.

As numbers, or biomass, and intensity of competition increase, a population comes to be not only more numerous but to contain increasingly more age/size groups. Eventually, the "asymptotic" population will have a number of different age groups and will closely resemble the complex age structure characteristic of unexploited fish populations before commercial fishing begins. Figure 6.3 indicates how such a population structure originates. After large-scale fishery operations have begun, we can expect a retrogression of the complex age/size structure towards the simpler, more "juvenile" condition of earlier stages of growth of the population (Huntsman, 1948; Cushing, 1968; Nikolskii, 1969, Chapters 5 and 7). The extent that the population structure becomes simplified will depend on the intensity and duration of fishing and

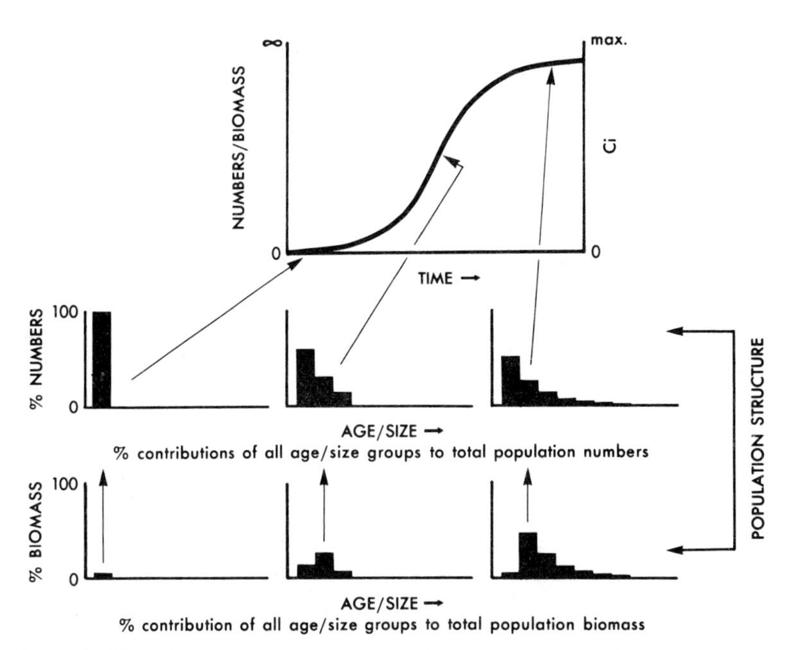

Fig. 6.3. This shows, semi-schematically, the changes in age/size structure of fish populations that may be expected to accompany the increase in numbers/biomass as the population grows. The absolute increase in numbers is shown to be represented by an age/size structure that gradually becomes more complex and in which young/small fish, though still proportionally dominant, are less dominant than they were originally. The important differences between this and the proportional contributions of the various ages and sizes to total biomass of the population are shown for comparison. Note (top graph) that *Ci* increases as population biomass increases. (Original.)

the minimum size for capture which is applied. Those individuals which remain in the population after fishing has begun will tend to grow more rapidly, females reaching breeding size earlier, therefore having less time in which to suffer mortality beforehand. These trends may offset, at least partially, the removal of larger female breeding stock—which anyway probably do not contribute very much to egg production in a "mature" population (see also Tables 6.1 to 6.3 and 6.6). Hence, the production of a more juvenile population structure through fishing may not necessarily be accompanied by a lowering of numbers or biomass to values similar to those in the growing population when it was young and had a simple age/size structure (Fig. 6.4; compare with Fig. 6.3).

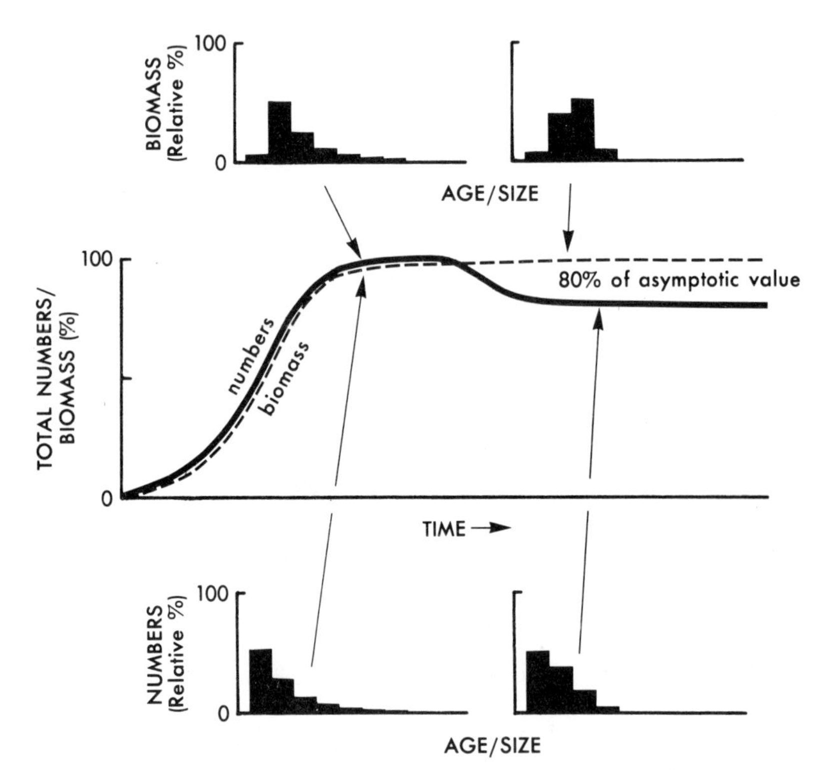

Fig. 6.4. Generalized scheme of the effects on fish population age/size structure as a result of fishing operations beginning on population that has grown to occupy an environment and is in "steady state". The upper and lower graphs to the left indicate the population age/size structure with respect to biomass and numbers at that time. The upper and lower graphs to the right indicate the trend of the changes to be expected after fishing has been in operation for a significant period. (Original, but drawing on Huntsman, 1948; Cushing, 1968; Nikolskii, 1969; and others.)

If, in a steady state population excluding emigration, numbers decline, either a factor causing mortality will have increased (e.g. fishing) or recruitment will have declined or both. If numbers are rising then mortality must be declining or recruitment will have become more successful or both. If a population is increasing, overall survival rate of individuals must be more than compensating for overall mortality. In a steady state population, number of deaths—over some regular period, probably a year—is nicely balanced by numbers entering the population through reproduction.

It is important to differentiate between the implications of growing towards an upper asymptote in the case of populations of those organisms which have relatively fixed individual growth rates and in fish, in which this is unusual. If a population of birds, mammals or insects exhibited a sigmoidal curve of growth the upper asymptote could have been determined by one or more of the following:

(i) A density-dependent effect on mortality rate which intensifies as population density mounts.

(ii) Inhibition or control of reproduction, also density-dependent, probably mediated through behaviour and sometimes resulting from the complete physical utilization of the reproductive substrate.

(iii) Toxic accumulation of metabolic wastes or substances inimical to reproduction, which grows with population.

II. STABLE AGE DISTRIBUTION, LIFE TABLES AND AGE-SPECIFIC FECUNDITY

The ecological importance of stable age distribution has been carefully stated by Andrewartha and Birch (1954), who emphasized (see also Chapter 2) that the true expression of the logistic curve of growth of numbers could be expected only in populations initially possessing a stable age distribution. I have employed the less precise but more comprehensive concept of "steady state" in connection with increasing fish populations for several reasons. Firstly, the growth of a population in a new space (referred to earlier) would normally follow not from the introduction of a large sample of an existing population, in which all the life stages were adequately represented in their final form as they would be in a stable-age population, but from the escape or colonizing entry of a relatively few individuals. These could be of any size or age as long as both males and females, reproductively mature or capable of becoming so, were represented. Moreover, if we observe an increase in a steady state population of fish, as a result of a more or less sudden improvement in food supply or increase in living space per fish, we expect initially a change in the *size* structure of the population as a result of an increase in the growth rate of individual fish. It would only be later, a "generation" later, that an increase in numbers through augmented recruitment could be expected. Whether, eventually, the age distribution that accompanied the original steady state might be expected to recur at higher levels of population has not been resolved for fish.

It has been assumed that growth in numbers and size of fish in populations would tend to be roughly sigmoidal, at least during early build up of population. But any suggestion that the curve of growth would be strictly logistic has been avoided for the reasons indicated.

Andrewartha and Birch (1954) discussed at length the application of life-tables in calculating age-specific fecundity. They stated that:

> Since the distribution of ages changes with time in all populations other than those in which the distribution is the stable age-distribution, the actual rate of increase will also change with time. This will be true even though the values for Lx and Mx (the birth-rates and death-rates in the different age groups) remain constant.

There is no doubt about the desirability of constructing life-tables and survivorship curves for fish populations where feasible. Nor should there be any reluctance to compute the age-specific (or better, size-specific) fecundity from such data. But whereas the sort of data quoted by Andrewartha and Birch (e.g. for flour weevils and voles) are based on animals whose age-specific fecundity is relatively unconfounded by the *size* of the adult, this is not so for fish whose labile growth patterns make the relation between age and fecundity uncertain. Fecundity in fish is apparently also related to feeding, hydrographic and other environmental conditions (Nikolskii, 1969, Chapters 3 and 4).

III. THE CAPACITY FOR INCREASE

The review by Braum (1967) appears to imply that there are considerable grounds indicative, though not really conclusive, of the fact that survival of larval fish is directly related to adequacy of food. There is, however, little or no certain evidence of density-dependent mortality among young (larval) fish, although various data cited by

Nikolskii (1969; especially Chapter 6) are interpreted by him as mortalities caused by relative inadequacy or abundance of the food supply. On the other hand, larval fish are generally vulnerable at such times as metamorphosis or when yolk sac absorption is coming to an end (Varley, 1967; see also Nikolskii, 1969). Furthermore, "general indications . . . are that there is no marked relationship between the abundance of the spawners and the number of subsequent recruits† within the range of population size for which data are available." Populations "seem to be inherently stable under the intensities of fishing which have so far been exerted . . . not . . . proceeding either to extinction or to an indefinitely greater size. We can take it that these populations are also stable in the virgin state" (Beverton and Holt, 1957). Such general observations suggest that a density-dependent mortality may act on young fish as a self-compensatory mechanism for ensuring similar rates of recruitment each year and as a buffer to offset change in number of spawners.

Beverton and Holt (1957) propose that in an equilibrium (steady state) population:

$$Ro = aEo \tag{6.1}$$

where Eo = eggs being laid each year, Ro = recruits surviving from them, a being a constant. If the population is to remain in equilibrium each group of recruits Ro must lay a total of Eo eggs during their lives, so

$$Eo = \gamma Ro \tag{6.2}$$

where γ = number of eggs laid per recruit.

If a sustained decrease in egg production per recruit occurs such that γ becomes γ', then

$$E = \gamma' Ro \tag{6.3}$$

These give rise to R_1 recruits so

$$R_1 = aE_1 \tag{6.4}$$

and these recruits to E_2 eggs where

$$E_2 = \gamma' R_1 \tag{6.5}$$

and the E_2 eggs to R_2 recruits where

$$R_2 = aE_2 \tag{6.6}$$

† Fish old enough to enter the area where fishing is in progress (Beverton and Holt, 1957).

From (6.3) and (6.4):

$$\frac{R_1}{Ro} = a\gamma' = \text{constant}$$

and from (6.5) and (6.6)

$$\frac{R_2}{R_1} = a\gamma' = \frac{R_1}{Ro}$$

so that, in general

$$\frac{Rn}{Ro} = (a\gamma')^n \tag{6.7}$$

This forms a geometric series among successive annual groups of recruits; since the ratio is less than unity ($a = 1/\gamma'$ and $\gamma' < \gamma$) the series is decreasing. Eventually such a situation could lead to population extinction, as indicated in Fig. 6.5. That this usually fails to happen in real populations can possibly be explained if "the mortality coefficient during one or more states of early life (is) influenced by the density of the larval fish themselves". Beverton and Holt give an expression for natural mortality of the post-recruit stage of the population

$$_M\left(\frac{dN}{dt}\right) = -MN \tag{6.8}$$

where N is the number of fish present and M is a constant mortality coefficient.

From this they propose that

$$_M\left(\frac{dN}{dt}\right) = \{\mu_1(t) + \mu_2(t)N\}N \tag{6.9}$$

where $\mu_1(t)$ is the density-independent component of mortality, whereas $\mu_2(t)$ depends on interaction between individuals.

Detailed development of this problem leads to the formula

$$R = \frac{1}{\alpha + \frac{\beta}{E}} \tag{6.10}$$

where R is the number of fish surviving to the beginning of the post-recruit phase (i.e. the actual number of recruits), E is the total number of eggs, α and β are constants for any given value of $t\rho$.†

† $t\rho$ is the time to reach the post-recruit stage in the life of fish in an exploited population (Beverton and Holt, 1957, p. 28).

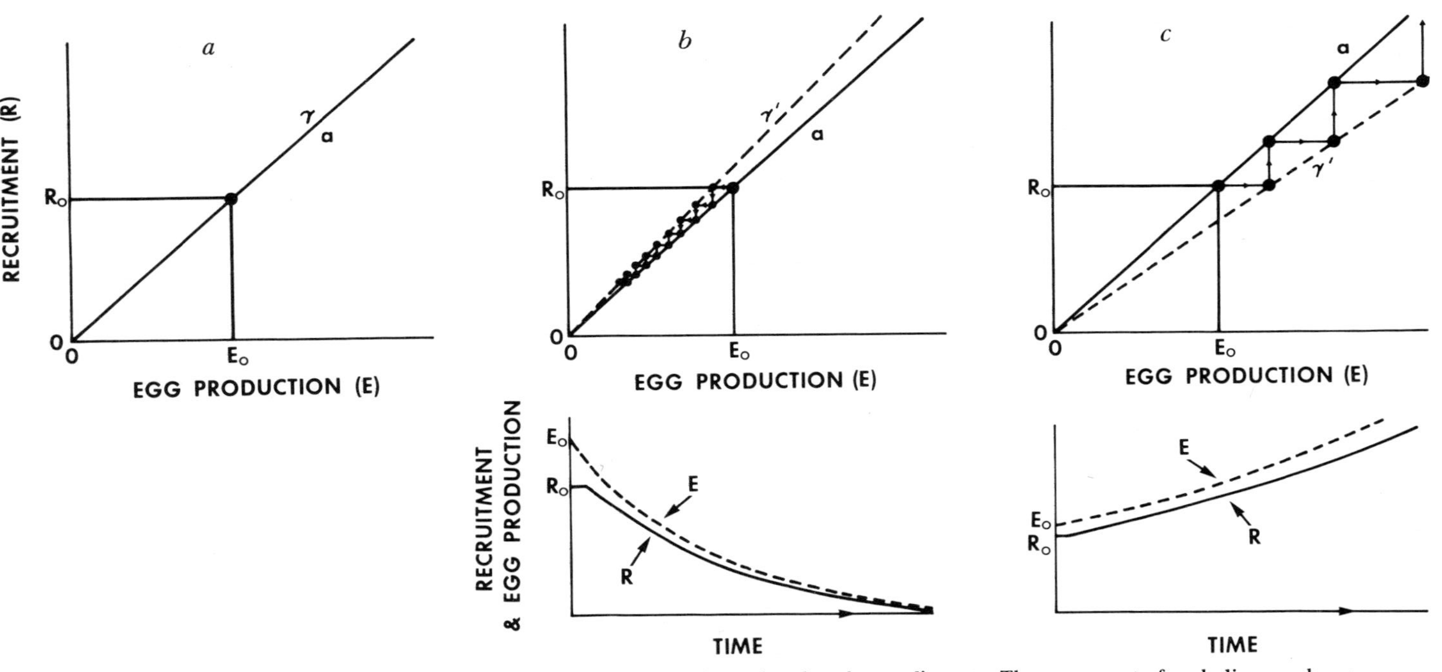

Fig. 6.5. The dynamics of a self-regenerating model with a density independent larval mortality rate. The upper part of each diagram shows successive values of recruitment and egg-production following a sustained change in egg-production per recruit of the adult population; in the lower part these are plotted as a time series. With this form of larval mortality a population cannot regain a steady state if a sustained change in egg-production per recruit occurs.

Fig. 6.5a. Egg-production and recruitment in the steady state.

Fig. 6.5b. Trends following a sustained decrease in egg-production per recruit.

Fig. 6.5c. Trends following a sustained increase in egg-production per recruit. (After Beverton and Holt, 1957.)

Beverton and Holt (1957) point out that an inherent property of this formula is that β is related only to μ, the coefficient of density-independent mortality, while α is related to the coefficient of density-dependent mortality as well. In demonstrating the implications of the formula they suggest the following data for a prerecruit phase of growth ($t\rho = 1{\cdot}2$ years), divided unequally into three stages with widely differing mortality coefficients for each of the three, thus:

Stage	Duration (years)	Mortality coefficients
1	0·3	$_1\mu_1 = 0{\cdot}827$
		$_1\mu_2 = 9{\cdot}113 \times 10^{-10}$
2	0·025	$_2\mu_1 = 19{\cdot}5$
		$_2\mu_2 = 0$
3	0·875	$_3\mu_1 = 1{\cdot}0$
		$_3\mu_2 = 5{\cdot}556 \times 10^{-10}$

Though this is a hypothetical example it is plausibly based on real situations. For instance, it is pointed out that $_2\mu_2 = 0$ might realistically represent the minimal density-dependent mortality during a short critical period (such as larval metamorphosis in a flatfish) when density-independent influences on mortality temporarily assume very high values.

Figure 6.6a shows the consequences of calculating from equation (6.10) pre-recruitment survival, shown in the three phases of the data table above, plotted against age for two hypothetical values of E. Two prominent features are the great drop in survival over a very short period (stage 2 in the data table) and the similarity in number of recruits despite marked dissimilarity of initial egg numbers. The latter is a consequence of an assumed density-dependence of larval mortality. As Beverton and Holt put it "as $E \to \infty$, R approaches a finite limit." After age $t\rho$ is reached mortality (and therefore recruit survival) is obtained from equation (6.8) above.

In Fig. 6.6b the general form of the relation between recruits and the egg number is shown in three curves calculated from equation (6.10) using the following data:

Curve (a) $\alpha_1 = 4 \times 10^{-9}$
$\beta_1 = 1$

Curve (b) $\alpha_2 = 1{\cdot}99 \times 10^{-9}$
$\beta_2 = 5{\cdot}005$

Curve (c) $\alpha_3 = 0{\cdot}5 \times 10^{-9}$
$\beta_3 = 20$

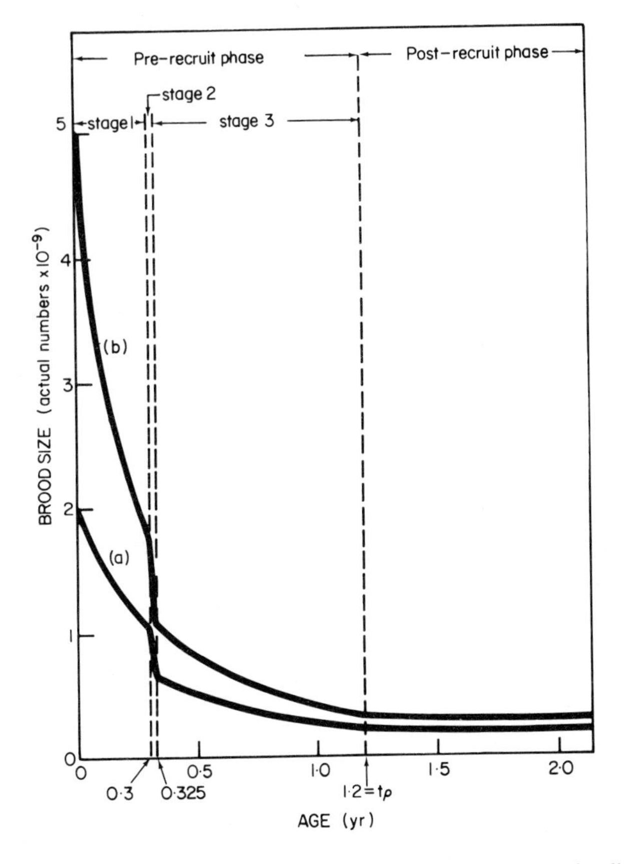

Fig. 6.6a. Pre-recruit survival curves with the larval mortality rate varying linearly with density. These are hypothetical examples in which the mortality rate is taken as changing abruptly at certain stages during the pre-recruit phase, as explained in the text. Note that the number of recruits (at age $t\rho$) are more similar in the two cases than are the number of eggs from which they survive; this is the result of making the larval mortality rate density dependent. (After Beverton and Holt, 1957.)

In (a), because of the very high value assigned to the coefficient of density-dependent mortality in larval fish, there is a situation where numbers of recruits cannot exceed a certain limiting value no matter how high the number of eggs. In curve (c) the effect of density is, by contrast, so slight that there is a nearly proportional change in R with E. Curve (b) represents a case intermediate between (a) and (c).

Figure 6.7 indicates the general outcome of density dependent larval mortality for a self-adjusting relationship of R and E. The straight line γ is identical to that which related R and E in Fig. 6.5,

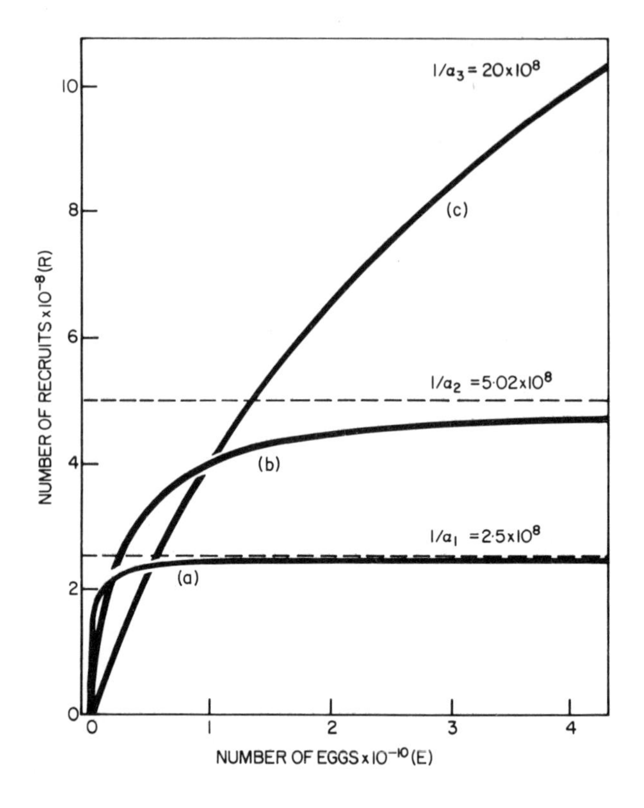

Fig. 6.6b. The relationship between egg-production and recruitment with a larval mortality rate varying linearly with density. The differences between the three curves are the result of varying the magnitudes of the parameters α and β of (6.10), thus altering the relative importance of the density independent and density dependent components of larval mortality. Curve (a) shows how this representation of larval mortality can result in recruitment being virtually independent of egg-production over a wide range—a feature which appears to be characteristic of many fish populations. (After Beverton and Holt, 1957.)

from equation (6.3), which showed that when larval mortality was not density-dependent, a population could not recover a steady state if subject to a sustained change in egg production. From Fig. 6.7 it is apparent that if larval mortality is density-dependent, a transitory change in egg production will lead to a recovery of stability. If the changes in egg production are sustained new steady states are produced.

Beverton and Holt (1957) also explored the implications for egg production and recruitment of density-dependent growth rate of larval fish. Figure 6.8 shows a hypothetical curve indicating that

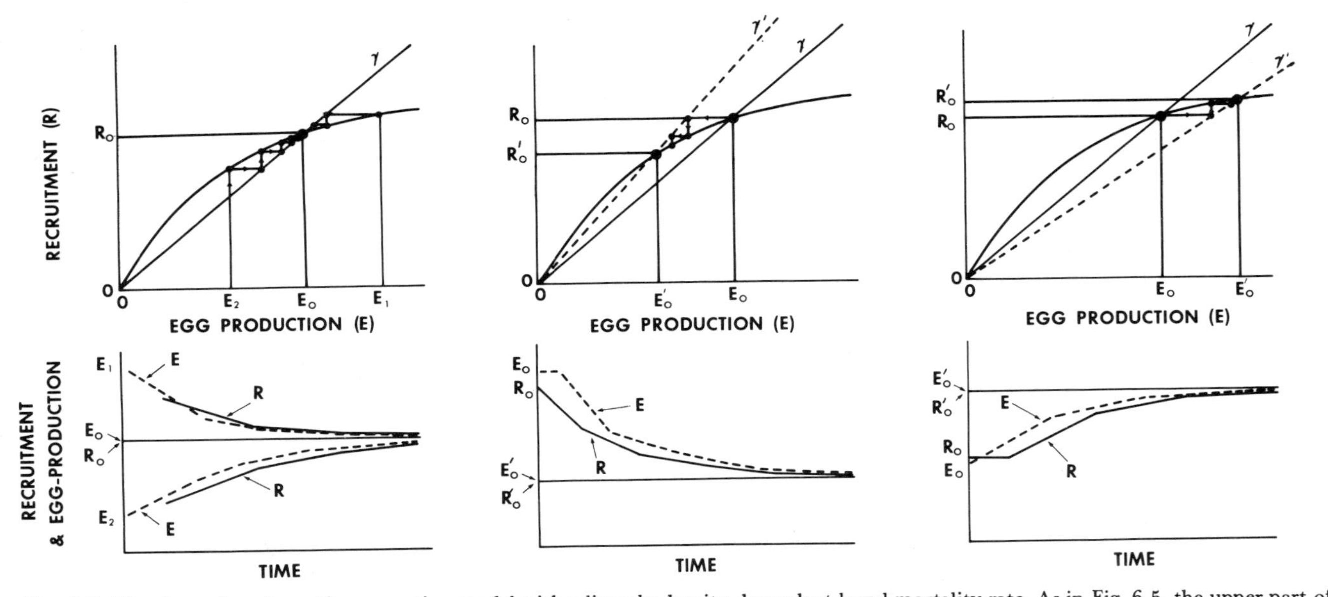

Fig. 6.7. The dynamics of a self-regenerating model with a linearly density dependent larval mortality rate. As in Fig. 6.5, the upper part of each diagram shows successive values of recruitment and egg-production following a change in egg-production per recruit of the adult population. In the lower part these are plotted as time series. With this form of larval mortality the population is "self-compensating"; after a transitory change in egg-production per recruitment (or in larval mortality) the original steady state is regained, while if the change is permanent a new steady state is reached. These properties should be contrasted with those shown in Figs 6.5 and 6.9.

Fig. 6.7a. Recovery of stability following a transitory change in egg-production.

Fig. 6.7b. Transition from one steady state to another following a sustained decrease in egg-production per recruit.

Fig. 6.7c. Transition from one steady state to another following a sustained increase in egg-production per recruit. (After Beverton and Holt, 1957.)

recruitment rises at first to a maximum as egg production increases, then declines again. The dynamics of self-regulation such a postulate leads to are indicated in Fig. 6.9. Depending on the relative slopes of the recruitment curves and egg production at their point of intersection "permanent" oscillations of egg production and recruitment can be set in motion, on disturbance of egg production even if

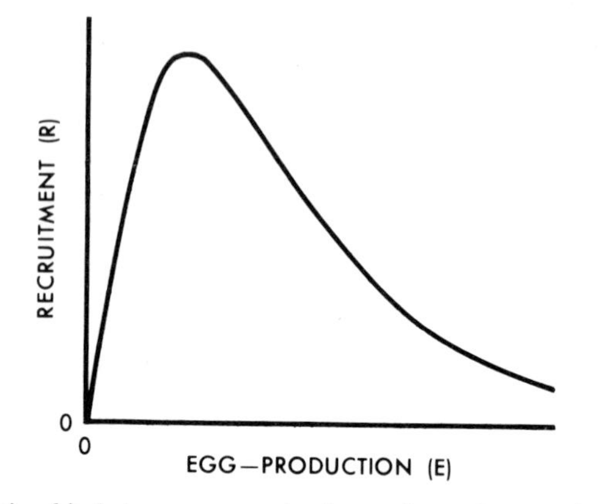

Fig. 6.8. A relationship between egg-production and recruitment when the growth rate of the larvae is density dependent. In this case a cause of larval mortality (e.g. predation) is assumed to operate while the larvae are within a limited size range. If growth of the larvae is density dependent the time they take to grow through this critical size range—and hence the severity of the mortality they suffer—increases as their density increases. This results in the egg-recruit curve reaching a maximum and the population having the kind of dynamic properties shown in Fig. 6.9. (After Beverton and Holt, 1957.)

the disturbance is transient. For these to appear, however, the disturbances must occur during steady states in which recruitment values lie to the right of the maximum value of the recruitment curve. Otherwise, the dynamics will be as in Fig. 6.7.

These models are all based on slender evidence of what actually occurs during egg and larval development in fish populations and the most that can be claimed is that Beverton and Holt (basing themselves largely on Ricker, 1954) have provided a series of potentially very useful procedures for examining possible outcomes of some postulated relationships between egg production and survival. However, these models relate only to egg and larval, and immediately post-larval, life. As Beverton and Holt admit, there is no clear indication that fish display density-dependent mortality in post-larval life.

It seems probable that in the majority of teleosts egg production is

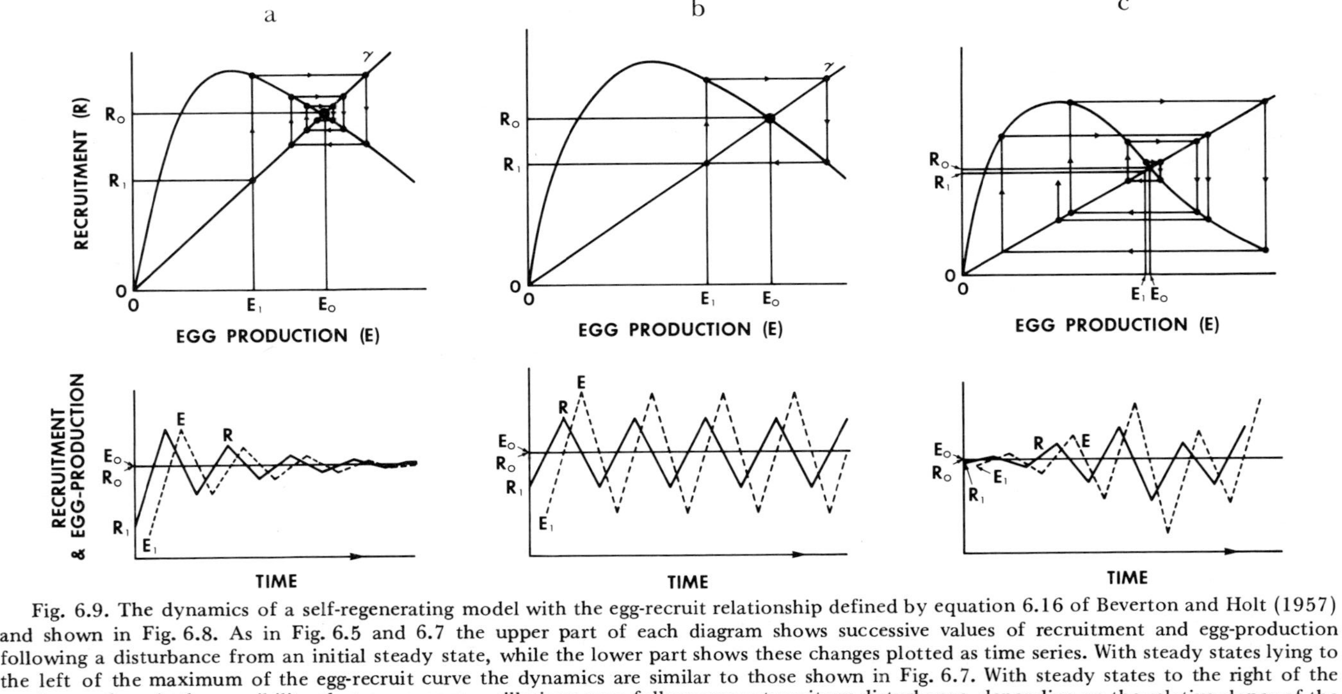

Fig. 6.9. The dynamics of a self-regenerating model with the egg-recruit relationship defined by equation 6.16 of Beverton and Holt (1957) and shown in Fig. 6.8. As in Fig. 6.5 and 6.7 the upper part of each diagram shows successive values of recruitment and egg-production following a disturbance from an initial steady state, while the lower part shows these changes plotted as time series. With steady states lying to the left of the maximum of the egg-recruit curve the dynamics are similar to those shown in Fig. 6.7. With steady states to the right of the maximum there is the possibility that permanent oscillations may follow even a transitory disturbance, depending on the relative slopes of the egg-production (γ) and recruitment curves in the region of their intersection.

Fig. 6.9a. Slope of recruitment curve shallower than that of egg-production curve at intersection. Recovery of initial steady state following a transitory disturbance.

Fig. 6.9b. Slope of recruitment curve equal to that of egg-production curve at intersection. Permanent oscillations of equal amplitude following a transitory disturbance.

Fig. 6.9c. Slope of recruitment curve steeper than that of egg-production curve at intersection. Permanent oscillations of increasing amplitude following a transitory disturbance. (After Beverton and Holt, 1957.)

closely related to size, either length or weight† (e.g. Nicholls, 1958c; Allen, 1951; Shapovalov and Taft, 1954; McFadden, 1961; Bagenal, 1967, 1969) so that the generalizations of Beverton and Holt (1957) for haddock and plaice populations may hold reasonably well for many other populations as well. Beverton and Holt also make the probably safe assumption that in species which release their ova for external fertilization, there is usually more than enough sperm. Their computations are therefore based only on the fecundity of *females*—in terms of eggs per unit of body weight. (Indeed, Nikolskii (1969) refers to many examples of highly variable sex ratios among fish which, it is claimed, are adaptive means for ensuring there are plenty of females when conditions are especially favourable for egg production.)

As we have seen (Fig. 6.9), Beverton and Holt included, among their generalized models, curves of the relationship between recruitment and egg production that have a dome shape—indicating that recruitment is highest at some "intermediate" value of egg production instead of increasing asymptotically with egg production. Herrington (1948) gave a very symmetrical sharp-domed curve to fit recruitment/egg production data for Georges Bank haddock. However, Beverton and Holt believed that most evidence suggested that an asymptotic curve was a more generally useful representation of the relationship for the majority of fish populations. It seems clear that what is really required is far more information on populations of different species of fish.

Ricker (1958) has shown that a family of reproduction curves relating mature progeny to spawners can be constructed, with forms ranging from asymptotic to various dome shapes. The relationship to which all conform is given as

$$Z = W e^{a(l-w)}$$

where $a = Pr/Pm$, $Z = F/P$ and $W = P/Pr$ and F is the filial generation (recruitment) at some stage after density-dependent mortality ceases; P is the parental generation; Pr is the "replacement" size of the parental generation, i.e. that which, on average, just replaces its own

† Nikolskii (1969; especially Chapter 3) cites many data which he believes demonstrate that neither egg number nor quality bear a simple or very constant relationship to size in fish populations. He considers that fish have a marked ability to adjust both these variables to the favourable situation of the food supply and other factors in the environment. But even if this view is accepted it seems unlikely to disagree seriously with the models of fecundity proposed below, although it would mean that considerable scope for intraspecific variability in size-specific fecundity and egg quality would have to be allowed for.

numbers; Pm is the level of parental stock which produces the maximum filial generation.

Ricker (1958) described methods of fitting this curve to actual data.

Ricker (1958, 1963) has also emphasized certain consequences of special shapes of reproduction curves in relation to fishing catches. He has (Ricker, 1963) given an especially interesting outline of the consequences to a fishery of a curve in which recruitment rises extremely rapidly at low adult stock densities to become independent of stock density over most of the range of the latter. Such a curve is related to a catch curve that rises smoothly and apparently asymptotically, but then drops precipitately beyond a certain rate of fishing—to zero yield. Ricker has warned that this tendency for rapid and excessive stock depletion beyond a certain fishing rate is not restricted to populations with flat-topped reproduction curves, but is also characteristic of populations in which the best fishing yield is being derived from a small breeding stock—i.e. where the optimum exploitation is removing a high proportion (say 75% or more) of the stock. In emphasizing that the precise management of a fishery can be a very exacting business, Ricker (1963) also points out that the variable nature of the recruitment is an added complication. In general Ricker (1963) advises management procedures that tend towards cautious rather than over optimistic assessments of catch potential.

Assuming that egg production for a population is approximately proportional to the total weight of mature females it contains (as in haddock or plaice populations), Beverton and Holt (1957) write

$$E = s\chi\eta\,\bar{P}w$$

where E is egg production of the population; $\eta\bar{P}w$ is the annual biomass of the mature section of the population; s is the percentage of mature females; χ is fecundity/g weight of females.

In fish populations inhabiting the sea, lakes, rivers and other large water bodies, models of egg production and recruitment such as those considered so far can be, at best, only approximations of reality for changes in population density, from whatever cause, may beset the post-larval population. Therefore even if density-dependent mortality among the eggs or young were able to bestow stability on the numbers entering the population as young-of-year, variations in survivorship patterns of older fish would still have to be contended with.

Let us for the time ignore the question of whether or not

mortality is density-dependent. Any factor changing the mortality patterns among post-larval fish will eventually modify the age or size structure of the population. An arrest or reduction of mortality would have the eventual effect of adding age groups not formerly present, while an overall increase in mortality could eventually reduce the number of age groups, thus simplifying the structure and making it more juvenile; these principles were mentioned earlier.

If, however, growth is density-dependent, or rather competition-dependent as Chapter 5 attempts to demonstrate, it may perhaps be seen as a means by which egg production in fish is controlled. If growth becomes depressed, in the present context, it would be assumed to result from food scarcity whether brought about initially by a high population density or by poor productivity of food organisms. Indeed, as far as fish growth responses are concerned the effects of food and space may probably be considered as having remarkably similar effects. Scarcity or plenitude of food are quite relative questions, depending on how much there is for all to share given the food needs for growth, maintenance, activity etc. of all fish in the population. If a fish's growth rate becomes reduced, because of competition for food, that could be viewed as a way of controlling the enlargement of the fish's reproductive system or perhaps even preventing it from functioning in certain extreme instances—for instance in Swingle's pond populations mentioned in Chapter 5.

The most important single effect of a maintained reduction in growth rate in fish may be to delay maturity, thus exposing the sub-adult fraction of the population to the causes of mortality for longer than usual and ensuring that fewer females will eventually reach sexual maturity (see Nikolskii, 1969; especially Chapter 4).

In those instances where food production increases, where a population is rapidly increasing to occupy a new environment or after an unusually high mortality, reverse principles to those described above may be expected to operate.

It is difficult to avoid the impression that growth flexibility is an adaptive mechanism by means of which fish populations tend always towards the same population age/size structure and the same individual growth rate, no matter what the nutritional status of their environment. To test this idea small numbers of young fish, or even of adult breeding stock of uniform size, could be released into various water bodies in which different amounts of food were available. So long as mortality rates were of the same order and pattern in the different populations and assuming the amount of spawning substrate was not limiting, it could be postulated that competition would control population. That is, in time, *Ci* would tend to reach a certain identical value for the various populations.

We turn now to the effects of growth of female fish on egg production. Figure 6.10 shows two different growth rates for females of a hypothetical species. These are the two rates that will be assumed in a series of simple computations of egg production in

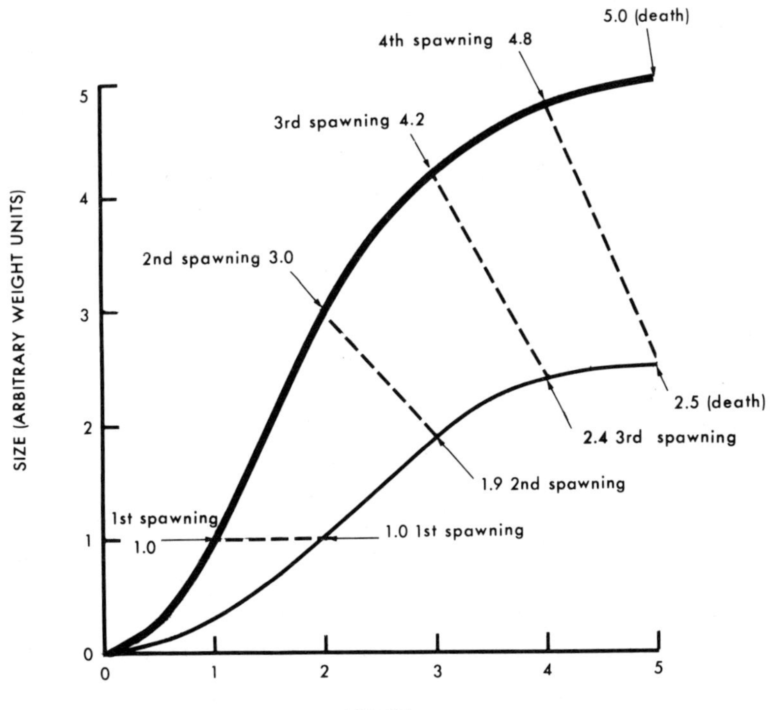

Fig. 6.10. Schematic size/age curves to indicate the effects of growth rate on *age* at first sexual maturity and *size* at all subsequent annual spawnings. "Fast" growth rate—heavy curve; "slow" growth rate—light curve. (Original.)

several imaginary populations. Assume that all the populations were begun in identical fashion by the placing of 2000 female† eggs in identical environments previously free of fish and that the populations grow in comparison pairs, one fast and one slow; the causes of differences in growth rate are not relevant here.

Again consider the population as not being in steady state but growing in a new space, a device useful in helping us understand how

† That is, eggs that will develop into female fish. It will be assumed that the same number of male eggs are present.

fish take advantage of a new situation and how they *must* do so because of the laws of growth, birth and death which rule their lives. This sort of model may frequently be overlooked in considering merely the "here and now" of steady state situations or such situations disturbed in specified regular ways (e.g. by changes in fishing). Of the four types of survivorship curve given in Fig. 6.1, only type III (Slobodkin, 1962) depicts "a system in which a constant fraction of the animals die at each age." Mean life expectancy at any age remains constant throughout the life of the animal. Many fish survivorship patterns may, for practical purposes, be considered as largely composed of an initial curve IV—but only the first part—i.e. all the descending early part of the curve, but little of the nearly horizontal limb after inflexion, then succeeded by (Slobodkin's) curve III. This type of composite curve is illustrated in Fig. 6.11 by two examples. The first shows a 95% mortality, the second a 99% mortality in the egg-larval stages. Both curves have the same survivorship pattern thereafter—50% each year—except during the fifth and last year of life during which, for convenience and to simplify the calculations, 100% mortality is supposed to occur. This treatment differs from Slobodkin's in that the curves should be considered as depicting the survivorship of *age groups* after Deevey (1947), whereas Slobodkin (1962) talks about the whole population which, as already indicated, will actually coincide with the survivorship curves of all age groups only in the case of a population with a stable-age distribution. Figure 6.12 gives the third set of data needed for the demonstration: a simple linear relation between egg number and fish size is assumed.

Tables 6.1 and 6.2 show the results of computing egg production and increase in population each year for two pairs of populations which all begin with 2000 female eggs. Of the two pairs, one pair is characterized by each of the mortality patterns given in Fig. 6.11. In addition, for one population of each pair, egg production is computed from the "fast" growth curve for females (Fig. 6.10) and in the other population, it is computed from the "slow" curve. An important consequence of this is that females of one population of each pair require a longer time to reach sexual maturity; see again Fig. 6.10.

The major effect of delaying the first breeding period for one year is apparent in both tables. Where 5% of the eggs survive to the post-larval stage the "slow" growing females are still unable to produce at nine years as many eggs as the "fast" growing ones at five years. This effect is even more enhanced in populations in which

Table 6.1

Egg production and increase in number of fish in two hypothetical populations characterized by different mean growth rates of their individuals; 95% mortality assumed during egg-larval development

Fast										
Year of Population Build-up		1	2	3	4	5	6	7	8	9
Egg Production (total)		2×10^{3}	5×10^{4}	$1\cdot275 \times 10^{6}$	$3\cdot378 \times 10^{7}$	$8\cdot935 \times 10^{8}$	$2\cdot364 \times 10^{10}$	—	—	—
No. of Post-larval Fish		10^{2}	$2\cdot5 \times 10^{3}$	$6\cdot375 \times 10^{4}$	$1\cdot689 \times 10^{6}$	$4\cdot467 \times 10^{7}$	$1\cdot182 \times 10^{8}$	—	—	—
	Year-group									
Totals of	0–1	50	$1\cdot25 \times 10^{3}$	$3\cdot185 \times 10^{4}$	$8\cdot444 \times 10^{5}$	$2\cdot234 \times 10^{7}$	$5\cdot910 \times 10^{7}$	—	—	—
age groups	1–2		25	$6\cdot25 \times 10^{2}$	$1\cdot593 \times 10^{4}$	$4\cdot222 \times 10^{5}$	$1\cdot117 \times 10^{7}$	—	—	—
surviving at	2–3		0	12·5	$3\cdot125 \times 10^{2}$	$7\cdot963 \times 10^{3}$	$2\cdot111 \times 10^{5}$	—	—	—
end of each	3–4		0	0	6·25	$1\cdot56 \times 10^{2}$	$3\cdot982 \times 10^{3}$	—	—	—
year	4–5		0	0	0	0	0	—	—	—
		Spawning								
Slow										
Year of Population Build-up		1	2	3	4	5	6	7	8	9
Egg Production (total)		2×10^{3}	0	$2\cdot5 \times 10^{4}$	$2\cdot375 \times 10^{4}$	$3\cdot275 \times 10^{5}$	$5\cdot934 \times 10^{5}$	$4\cdot564 \times 10^{6}$	$1\cdot149 \times 10^{7}$	$2\cdot022 \times 10^{8}$
No. of Post-larval Fish		10^{2}	0	$1\cdot25 \times 10^{3}$	$1\cdot188 \times 10^{3}$	$1\cdot638 \times 10^{4}$	$2\cdot967 \times 10^{4}$	$2\cdot282 \times 10^{5}$	$5\cdot743 \times 10^{5}$	$1\cdot011 \times 10^{7}$
	Year-group									
Total of	0–1	50	0	$6\cdot25 \times 10^{2}$	$5\cdot94 \times 10^{2}$	$8\cdot188 \times 10^{3}$	$1\cdot484 \times 10^{4}$	$1\cdot141 \times 10^{5}$	$2\cdot872 \times 10^{5}$	$1\cdot664 \times 10^{6}$
age groups	1–2		25	0	$3\cdot125 \times 10^{2}$	$2\cdot97 \times 10^{2}$	$4\cdot094 \times 10^{3}$	$7\cdot418 \times 10^{3}$	$5\cdot705 \times 10^{4}$	$1\cdot436 \times 10^{5}$
surviving at	2–3		0	12·5	6·25	$1\cdot56 \times 10^{2}$	$1\cdot49 \times 10^{2}$	$2\cdot047 \times 10^{3}$	$3\cdot709 \times 10^{3}$	$2\cdot853 \times 10^{4}$
end of each	3–4		0	0	0	0	78	75	$1\cdot024 \times 10^{3}$	$1\cdot854 \times 10^{3}$
year	4–5		0	0	0	0	0	0	0	0
			Spawning							

Table 6.2

Egg production and increase in number of fish in two hypothetical populations characterized by different mean growth rates of their individuals; 99% mortality assumed during egg-larval development

Fast		1	2	3	4	5	6	7	8
Year of Population Build-up		1	2	3	4	5	6	7	8
Egg Production (total)		2×10^3	1×10^4	$6{\cdot}5 \times 10^4$	$4{\cdot}605 \times 10^5$	$2{\cdot}925 \times 10^6$	$1{\cdot}851 \times 10^7$	—	—
No. of Post-larval Fish		20	10^2	$6{\cdot}5 \times 10^2$	$4{\cdot}605 \times 10^3$	$2{\cdot}925 \times 10^4$	$1{\cdot}851 \times 10^5$	—	—
	Year-group								
Totals of	0–1	10	50	$3{\cdot}75 \times 10^2$	$2{\cdot}303 \times 10^3$	$1{\cdot}463 \times 10^4$	$9{\cdot}255 \times 10^4$	—	—
age groups	1–2	0	5	25	$1{\cdot}88 \times 10^2$	$1{\cdot}152 \times 10^3$	$7{\cdot}315 \times 10^3$	—	—
surviving at	2–3	0	0	2·5	12·5	94	$5{\cdot}76 \times 10^2$	—	—
end of each	3–4	0	0	0	1·25	6·25	47	—	—
year	4–5	0	0	0	0	0	0	—	—
		Spawning							
Slow									
Year of Population Build-up		1	2	3	4	5	6	7	8
Egg Production (total)		2×10^3	0	5×10^3	$4{\cdot}75 \times 10^3$	$1{\cdot}55 \times 10^4$	$2{\cdot}388 \times 10^4$	$4{\cdot}784 \times 10^4$	$1{\cdot}038 \times 10^5$
No. of Post-larval Fish		20	0	50	48	$1{\cdot}55 \times 10^2$	$2{\cdot}39 \times 10^2$	$4{\cdot}78 \times 10^2$	$1{\cdot}038 \times 10^3$
	Year-group								
Totals of	0–1	10	5	25	24	78	$1{\cdot}19 \times 10^2$	$2{\cdot}39 \times 10^2$	$5{\cdot}19 \times 10^2$
age groups	1–2	0	0	2·5	12·5	12	39	59·5	$1{\cdot}20 \times 10^2$
surviving at	2–3	0	0	0	1·25	6·25	6	19·5	30
end of each	3–4	0	0	0	0	0	3·1	3	9
year	4–5	0	0	0	0	0	0	0	0
			Spawning						

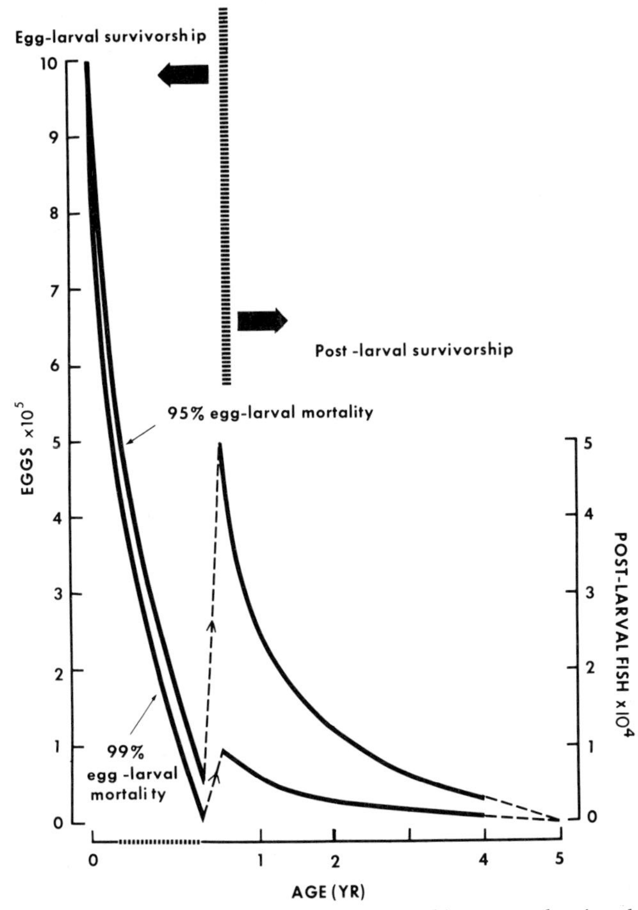

Fig. 6.11. Two schematic but semi-realistic survivorship curves showing the effects of 95% and 99% mortality during the egg-larval phase of life on the survival thereafter. Effects are important in relation to a cohort's ability to produce eggs; see text and Tables 6.1 to 6.6 for further discussion. (Original.)

only 1% survive to post-larval stages, when "fast" growers take only four years to surpass what the "slow" growers need eight years to produce.

The effect of higher mortality in the egg-larval stages also ensures that egg production in either "fast" or "slow" populations is greatly exceeded by the corresponding populations with lower egg-larval mortality (compare Table 6.1 with Table 6.2). On the other hand, when the "fast" population with high (99%) egg-larval mortality is compared with the "slow" population with low (95%) egg-larval mortality, it is seen that the former readily surpasses the latter in egg

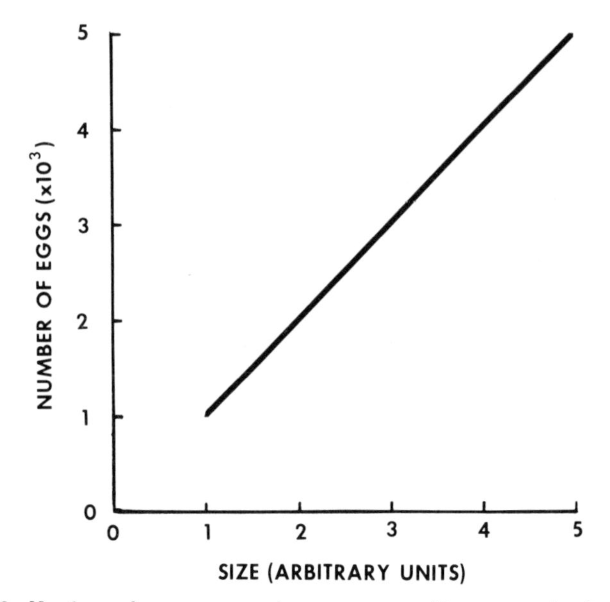

Fig. 6.12. Number of eggs versus size—a curve to illustrate a simple hypothetical relationship for computation of cohort egg-production; see text and Tables 6.1 to 6.6 for further discussion. (Original.)

production. Thus while the rate of larval mortality is certainly significant, that of growth is perhaps even more so. Pattern and rate of growth largely determine not only age at first maturity, but continue to influence the egg productivity of every successive age group; egg productivity being a function of size rather than age.

Nikolskii (1969) might oppose the above models on two grounds: first that size at first maturity among females is not constant, second that females do not always breed every year (or with seasonal regularity). However, even Nikolskii's own work does indicate a considerable tendency for first maturity to be more closely related to size than age. As to the second point, there is no particular problem about allowing for irregular or non-annual spawning in the models or, for that matter, spawning that occurs more frequently than annually.

Figure 6.13 features three hypothetical survivorship curves for post-larval fish. Curve A is of a type similar to type III shown in Fig. 6.1 which signifies a risk of death constant at all ages, but Curve A in Fig. 6.13 represents 80% mortality in the first year (post-larval) and 50% mortality thereafter—except after the fourth birthday, since for simplicity it is assumed no parent fish survive beyond five years. Curve B is the same as curve II of both Deevey

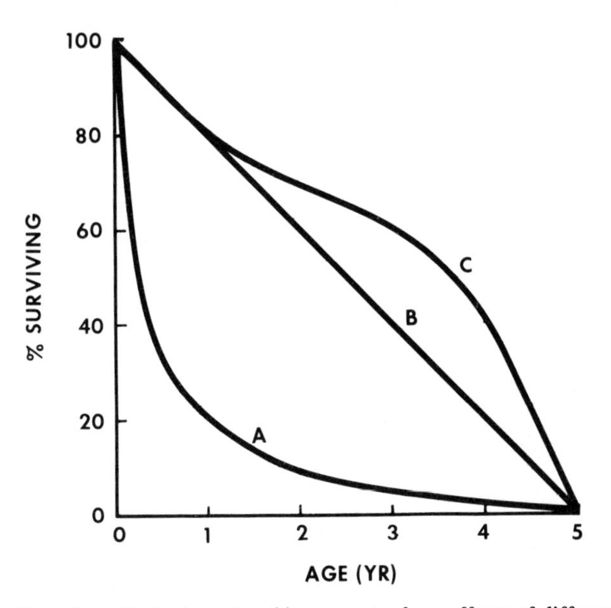

Fig. 6.13. Three hypothetical survivorship curves to show effects of different patterns of cohort survival on egg-production. Curve A represents 80% mortality in first year, 50% annual mortality thereafter; curve B signifies constant number of deaths annually; curve C has low early mortality, even less in "middle life", but survivorship declines sharply later. These curves are used in Tables 6.3 to 6.5. See also text for further discussion. (Original.)

(1947) and Slobodkin (1962) which signifies a constant number of deaths per unit time (see Fig. 6.1). Curve C combines features of types II and I, shown in Fig. 6.1, i.e. early mortality is relatively low, in "middle life" becomes even less then steepens late in life Curve C is a rather unlikely shape for a fish survivorship curve except for types such as salmon and eels which do not, in any case, spawn on a regular seasonal basis. However, it is included for general comparative purposes.

Table 6.3 shows egg production in each year of the life of a cohort of fish, using the growth curves of Fig. 6.10, egg production data of Fig. 6.12 and survivorship data of Fig. 6.13.

Starting from an assumed population cohort of 1000 post-larval female fish in each instance we can examine the effects of growth rate of females and of different patterns of mortality on egg production by year-classes. Considering the data calculated from curve A first, we see that the earlier years of the life of a year-class are those of the greatest egg-productivity. In the case of the "fast" population the second spawning produces most eggs. In the "slow" population, there is little difference between the egg productivities of

Table 6.3

Egg production in each year of the life of an original cohort of 1000 post-larval fish. Values calculated from growth curves of Fig. 6.10, egg/size data of Fig. 6.12, survivorship curves of Fig. 6.13.

Spawning		Size of Fish	Eggs per Fish	Progressive No. of Fish According to Survivorship Curve			No. of Eggs for Each Age Group According to Survivorship Curve		
				A	B	C	A	B	C
Fast-growing	1	1·0	1 000	200	800	800	200 000	800 000	800 000
	2	3·0	3 000	100	600	700	300 000	1 800 000	2 100 000
	3	4·2	4 200	50	400	600	210 000	1 680 000	2 520 000
	4	4·8	4 800	25	200	400	120 000	960 000	1 920 000
Totals							830 000	5 240 000	7 340 000
Slow-growing	1	1·0	1 000	100	600	700	100 000	600 000	700 000
	2	1·9	1 900	50	400	600	95 000	760 000	1 140 000
	3	2·4	2 400	25	200	400	60 000	480 000	960 000
Totals							255 000	1 840 000	2 800 000

first spawning and second spawning fish. In both "fast" and "slow" populations the last spawning of life produces considerably fewer eggs than the more prolific earlier ones. In those populations (both "fast" and "slow") showing the B type of survivorship pattern, there is greater egg productivity in the second spawning than before or after. The populations featuring the C type of survivorship also have their major egg productivity somewhat later (indeed, in the third spawning in the "fast" populations).

These three sets of data provide several lessons. For instance, the increase in size of individuals as they grow older is very significant because it means that as fish grow older they do not necessarily lose all importance in their capacity to contribute to the population's total egg productivity, just because of the effect of cumulative mortality on their particular year-class. And this obviously holds true for a considerable range of growth rates and survivorship patterns.

However, it must also be admitted (Tables 6.4 and 6.5) that in growing populations the *proportional* contribution of older year-classes to total egg productivity of the whole population of spawners is likely to be quite small, *despite* the relatively large size and therefore high egg productivity of such individuals. This is entirely because there are so few of them. This is also illustrated in actual populations in steady state, in which the proportional contribution of the older (larger) fish may, under ordinary circumstances, be very small indeed. Table 6.6 is constructed from data of Nicholls (1958a) on trout in the north-west rivers of Tasmania. It is based on survivorships drawn from his catch curve† (Ricker, 1948) from anglers' returns over a period of five years combined with the average length-for-age of trout from the freshwater sections of north-west rivers and data from graphs of egg production for fish of different lengths (Nicholls, 1958c). Some approximations are needed to adjust for the varying sources of these data (for instance, the data on fecundity were derived from lake fish in central Tasmania and river fish from the south). The assumption is also made that all females of

† Catch curves (see also Nikolskii, 1969; especially Chapter 6) are frequently the nearest approximation, available to fishery biologists, of survivorship curves. The "descending limb" of a catch curve indicates the proportional representation of various age/size groups, in a fish population, that have reached the size where no members of the group are too small to escape the gear used in capture. A catch curve also has an "ascending limb" which does not give proportional representation of the age/size groups it covers because numbers ranging from all of them to only a few are too small to be captured by the gear or method employed. Obviously, catch curves cannot usually reveal absolute numbers of fish present in a population and if the "descending limb" is to be used as an approximate survivorship curve this is valid only if the population can be shown to be in a steady state.

Table 6.4

Contribution (%) to total egg productivity of whole spawning population of all year classes as population grows; 95% mortality assumed during egg-larval development

Fast										
Year of Population Build-up	1	2	3	4	5	6	7	8	9	10
Year-group										
0–1	[a]	100	98·04	94·29	94·50	94·50	—	—	—	—
1–2	0	0	1·96	5·55	5·35	5·36	—	—	—	—
2–3	0	0	0	0·16	0·15	0·14	—	—	—	—
3–4	0	0	0	0	insignif.	insignif.	—	—	—	—
4–5	0	0	0	0	0	0	—	—	—	—
	Spawning									
Slow										
0–1	[a]	0	0	0	0	0	0	0	0	0
1–2	0	0	100	0	95·42	50·05	89·70	64·58	85·72	71·00
2–3	0	0	0	100	0	49·95	6·20	33·86	10·59	26·80
3–4	0	0	0	0	4·58	0	4·10	1·56	3·69	2·20
4–5	0	0	0	0	0	0	0	0	0	0
		Spawning								

[a] 2000 eggs planted.

Table 6.5

Contribution (%) to total egg production of whole spawning population of all year classes as population grows. 99% mortality assumed during egg-larval development

Fast									
Year of Population Build-up	1	2	3	4	5	6	7	8	9
Year-group									
0–1	[a]	100	76·92	81·43	78·72	—	—	—	—
1–2	0	0	23·08	16·29	19·28	—	—	—	—
2–3	0	0	0	2·28	1·79	—	—	—	—
3–4	0	0	0	0	0·21	—	—	—	—
4–5	0	0	0	0	0	—	—	—	—
	Spawning								

Slow									
0–1	[a]	0	0	0	0	0	0	0	0
1–2	0	0	100	0	83·3	50·26	67·43	57·35	—
2–3	0	0	0	100	0	49·74	19·71	35·71	—
3–4	0	0	0	0	17·7	0	12·86	6·94	—
4–5	0	0	0	0	0	0	0	0	—
		Spawning							

[a] 2000 eggs planted.

Table 6.6

Life table data for trout populations from the north-west rivers of Tasmania to show age/size specific fecundity and the contribution of the various age/size groups to the total egg production. Derived from the data of Nicholls (1958a, c).

Age (yr)	Mean Length-for-Age (cm)	Number of Fish in Population (%)	Eggs per Fish (E)	Proportion Total No. Eggs Laid (%)
2	23	80·5	500	68·3
3	29	14·9	800	20·3
4	38	3·8	1400	9·0
5	48	0·8	1800	2·3
6	52	0·04	2100	0·1

reproductive age breed successfully every year. However these are minor points and need not detract from the general significance of the table, which shows that because of the relatively high mortality per year the older fish do not, in this fishery, contribute much to the egg production, despite their large size and high individual fecundity.

However, it should be clear that all cases of growth and survivorship can be separately and accurately computed for their potential contribution to population growth and egg production.

Generally, populations selectively exploited by man or other predators will, as a result, tend towards more rapid growth (as mentioned earlier). Though the breeding stock may be affected adversely by predation if the size (age) at capture be small enough, there are in fish populations notable safeguards against the effects of removal of breeders. For instance, Table 6.3 should now be reread as though its data were obtained from established steady state populations. The "number of eggs" column now gives the egg productivity of each year group, assuming general patterns of survivorship do not change in any of the six populations considered and basic recruitment into the population each year remains stable. The proportional contribution of eggs by even the oldest age groups is still quite considerable.

If we take the example of expanding populations (Tables 6.1 and 6.2) then, even after a number of years, old (but therefore large) fish can still produce a high absolute number of eggs although relatively few such fish remain in the population. The biological implications of this are as follows. Suppose, for instance, that a population's breeding activities were reduced or even eliminated for one or more years, so that no young, or very few, were produced. That might occur as a result of siltation, poisoning or other destruction of breeding areas (for example by flooding), the effects of a pathogen or an efficient predator on the eggs or through behavioural disruption of spawning. While the proximate cause is of no immediate concern the effect of little or no recruitment in the presence of an otherwise normal survivorship pattern would produce a population structure deficient in young (small) individuals. Although the older spawners would be much less numerous than the younger spawning classes usually were and although this would be especially noticeable if the failure of the population to spawn continued for more than a season, their relatively large size should still ensure that they could produce sufficient eggs to start the population on its upward climb again. Besides, in many populations, the effect of an absence of recruitment would be to reduce the competition for food among the remaining fish, thus they would grow more rapidly, be larger for their age and produce more eggs than before. (Note that Nikolskii (1969; especially Chapter 3) doubts that annually reproducing fish can produce eggs at all beyond a certain age or size.)

It could be objected that too much attention has been given to capacity for increase and ability to recover from numerical setback when, after all, most populations are not in process of violent change. It is, however, properties of numerical increase combined with the pattern and magnitude of growth of individuals that bestow on fish populations not only their capacity for increase but their ability to maintain themselves dynamically. The individual animal may display fully its capacity for growth in the young, proliferating population or less fully, but perhaps as importantly, in its role in the established population which has been in steady state but which suffers a change.

If we turn again to Fig. 6.2 it should be apparent that, although years of active growth of individuals and of numbers may have preceded the establishment of the upper asymptote of population numbers, we can tell very little about the maintenance of that value by mere inspection or routine sampling of the population. For instance the population represented by the upper asymptote could consist of fish in which individual growth was almost negligible—precisely the sort of situation encountered in Swingle's pond fish populations in the last chapter. Such a population might be capable of reproducing, but high competition for food and consequent slow growth of individuals would tend to suppress the size of spawning females at every age.

Of three main sorts of determinants of the upper asymptote to numbers in populations as mentioned earlier for birds, mammals and insects, the first two (density-dependent mortality and toxic accumulation of wastes) will usually be of little significance to fish. The most direct effect of competition for food—death for the less successful—is moderated for fish because of their marked abilities to adjust growth and activity to availability of food and their capacity, if called upon, to "degrow" for extended periods. They seem to have substituted the mechanism of competition-dependent growth for that of density-dependent mortality. The subtle and complex interactions between mortality, growth and reproductive activity among fish make it very hazardous to compare different populations. Because of favourable opportunities for reproduction and good survival there are fish populations in which a large number of small (even "stunted") individuals is the rule (Nikolskii, 1969). And because facilities for spawning are sometimes very poor we can just as readily find populations in which there are relatively few, well-grown individuals (Lake, 1957). The biomasses of the two sorts of populations or at least the food supplies needed to maintain both kinds could, however, be very similar.

IV. GROWTH OF POPULATIONS AND INDIVIDUALS COMPARED

In searching for analogies between growth of individual fish and of populations of fish we can view the problem under two heads: growth curves and growth compensatory phenomena.

A. Growth Curves

Sigmoidal growth curves help considerably in characterizing what may be termed the general pattern or trend of growth whether of an individual or an organized group of individuals (population). However, the curve of growth in numbers of a fish population is smooth only as a first approximation. On magnification it will be found to be composed of many sub-curves which signify the influence of continual changes in mortality and recruitment. In this it will resemble the summation curves of growth of individual fish, which will usually vary seasonally, with food supply and with changes in competition for food (Chapters 2 and 4).

B. Growth Compensatory Phenomena

In the individual, overall growth in size indicates that additive processes are exceeding subtractive processes. The individual in steady state will merely maintain bulk and form which, as indicated in earlier chapters, means that processes of repair and maintenance of tissues are just sufficient to compensate for losses due to cell death, metabolism etc. There is a considerable analogy between these processes and a population in steady state. In the latter, new individuals entering the population nicely balance the total mortality rate in all age/size groups. If the recruitment was continuous, as it could be if fish bred throughout the year and mortality rate was unvarying, the metabolic analogy with an individual in steady state would be particularly close. As it is, steady state populations frequently vary in a regular yearly cycle. They can best be compared, therefore, with individuals which show seasonal changes in weights without corresponding change in linear dimensions (i.e. condition) or seasonal cycles of tissue destruction and repair, activity, excretion and spawning, which do not ultimately result in progress in size. If part of an individual be destroyed, considerable replacement and repair is often possible at a measurable cost in terms of matter or its energy "equivalent". If, in a steady state population, mortality be increased, there will be compensatory growth on the part of the remaining individuals, because of reduced competition for food. By the same token, individuals will increase in bulk if they receive more food and the same will occur in the biomass of a population whose nutritional status is elevated. Initially, biomass will increase because of a general increase in the growth rate of individuals; this will ensure a larger size of breeding females, which will produce more eggs. And in the course of a single generation an increase in numbers may follow. This will, in turn, have the effect of again reducing mean growth rate. A final balance will be struck which reflects the new food supply in terms of a higher biomass, even though the growth characteristics of the individuals in the population eventually may not differ greatly from those before the increase in food.

To keep a metabolic balance sheet on growth processes in populations many of the same principles as for individuals are applicable. And the fish that die in the course of growth must be taken into account. The principles are given more fully in Chapter 10.

7 | Predator-Prey Relationships Among Fish

Various reports testify to the predatory abilities of many species of fish but, unfortunately, few recorded observations are of direct use to population ecologists, because the majority were made without adequate accompanying evaluations of predator-prey relationships. The ecologist needs to know the kind, number and size of prey a predator consumes and although there may often be considerable information on these points, it is also very important to know how frequently and abundantly the predator feeds; such knowledge is generally much less accessible. An understanding of structural and functional specializations of fish for feeding also helps ecologists to postulate the exact nutritional role (niche) of predators, so that the terms in which the quest for food proceeds within trophic systems may be more fully characterized (see Chapter 5 and Nikolskii, 1963, 1969; Greenwood, 1963; Marshall, N. B., 1965; Popova, 1967).

With such observational data secured the investigator can turn to the next major problem in predation: the physiological rate of utilization of prey as food and the bearing of this on growth and maintenance of both predator and prey populations. Though there are many deficiencies in this area, some relevant information exists (Beverton and Holt, 1957; Paloheimo and Dickie, 1965, 1966a, b; Nikolskii, 1963; Ivlev, 1961; Allen, 1951; Ricker, 1954), and the guide lines for future studies have been established.

The major deficiency in the understanding of predator-prey systems among fish is in relation to population dynamics (Popova, 1967; Nikolskii, 1963, 1969).

It is difficult to understand why this subject has been so neglected though it is obvious that most of the thoughts of biologists have been given to the dynamics of fishing, itself a rather specialized form of predation on fish. Fishing is considered only as it relates fairly directly to the main themes in this book, as in Chapter 6 when it was necessary to consider population age/size structure and how this could be affected by selective removal of population members; but see the discussion below.

In this chapter we consider some experiments on prey-predator systems in ponds and outline the possible significance of such systems in natural populations. A model of prey-predator relationships is then discussed and its implications traced in relation to other work. The particular case of pike as predators is then mentioned and some other aspects of the topic are described.

I. CASE STUDIES OF FISH PREDATOR-PREY SYSTEMS

A. Swingle's Work on Pond Fish

We turn again to the studies of H. S. Swingle for evidence about predator-prey relationships in fish populations. These results, largely neglected by most ecologists, remain almost the only ones on a field scale repeated frequently enough to demonstrate, with something approaching certainty, the character of predator-prey relations among fish. The systems maintained by Swingle were admittedly simple ones, but less simple than many artificial laboratory systems which have been held to provide convincing evidence of the nature of predation (Ivlev, 1961).

Swingle's interest in predators and prey among fish stemmed from his need to control excessive reproduction of fish such as bluegills in farm ponds. His interest was practical rather than fundamental or theoretical, which helps to explain the relative neglect of his findings.

Swingle already knew that individuals in wild bluegill populations sometimes grow quite rapidly. Yet it seemed obvious that growth cessation under crowded conditions in chemically enriched ponds could not be attributed to low food productivity, because small, so-called "stunted", fish were numerically abundant. Their total biomass could also be quite high, as high as in populations in which fish were growing well. Swingle reasoned that sometimes fish were unable to grow satisfactorily because food supply *per fish* was inadequate, an inference which perhaps appears extremely obvious to fish ecologists nowadays. Thirty years ago it was not so obvious.

Swingle discovered that bluegill fingerlings stocked in ponds at 400 per ac (approximately 1000 per hec) grew rapidly until they reproduced, when the rapid growth of approximately 4000 young produced per female soon stopped the growth of all fish in the pond through competition for food. Swingle perceived that a suitable predator might consume enough young bluegills (and enough of its own young) to produce a "balanced"† population.

In early experiments in association with E. V. Smith, Swingle used bluegills as the basic "forage" species in ponds and attempted to control their increase in numbers by adding suitable proportions of predator fish (Swingle and Smith, 1940). In a typical early experiment an average-sized (1·3 ac), fertilized pond was stocked with bluegills and largemouth bass fingerlings during the late winter, at rates of 1500 and 100 per ac (=3705 and 247 per hec) respectively. Most of the bluegills originally released were still surviving after one growing season and were then classified as large (mean weight 118 g). These comprised 90% of the weight of the bluegill population, the remainder (6320 per ac = 15 610 per hec) being young-of-the-year. Bass consumed most of the young bluegills and also ate most of their own young, reducing them from the thousands which hatched during summer to a mere 141 per ac, a number roughly comparable to the number of bass originally released as fingerlings.

Even earlier experiments by Swingle and Smith (1940) had demonstrated that other predatory species (such as white crappy) though they fed voraciously on bluegills, were relatively ineffective in reducing numbers of the young of that species to the point where growth could again proceed. Nor could they adequately control the numbers of their own young.

Later Swingle (1949) published one of his most important studies of predation and "balance" in populations, in which he described numerous experiments using species such as golden shiner *Notemigonus crysoleucas,* gizzard shad *Dorosoma cepedianum,* goldfish *Carassius auratus,* mosquito fish *Gambusia affinis* and bluegills as the forage fish in combination with largemouth bass as the predator. In these experiments Swingle was testing several ideas. He was dissatisfied with bluegills as pond fish because of their spawning period: "Bluegills spawn in central Alabama from the last of May to the first of October. Consequently young bluegills for bass food are scarce during the period from January to June." Swingle was hoping to find suitable "forage" fish among species that would spawn at times when bluegills were not producing many young.

† In Swingle's (1950) terms, "balanced" populations "yield, year after year, crops of harvestable fish that are satisfactory in amount when the basic facilities of the bodies of water containing these populations are considered."

Combinations of these with bluegills might then yield a more continuous supply of young prey fish for bass during more months of the year. Golden shiners, which spawned heavily only from late March to early April (spring), though they did produce additional young irregularly throughout the summer period until late October, were not able to maintain their numbers in fertilized ponds that also contained bass (Swingle, 1946). Their slow growth rate ensured a

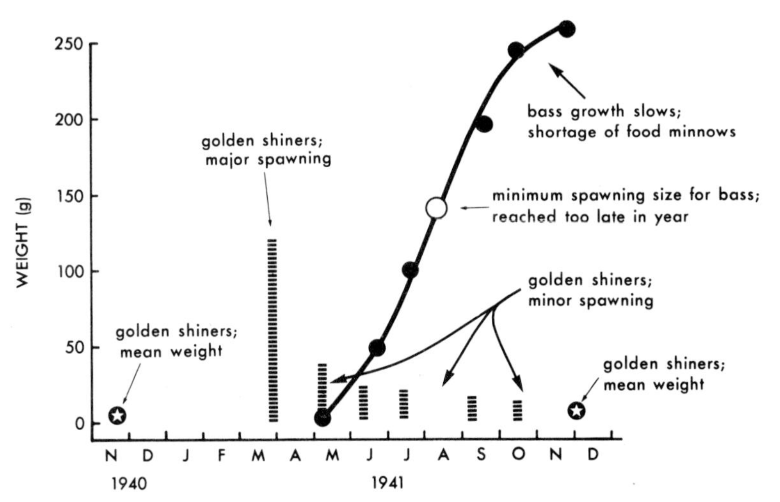

Fig. 7.1. Growth of largemouth bass as predators in fish pond in combination with golden shiners as prey. See text for further explanation. (Constructed from the data of Swingle, 1946.)

lengthy exposure to predation by bass before they became too big to be eaten. Their relatively low rate of reproduction in summer after an early spring maximum led, in combination with the slow growth, to a diminishing brood stock and eventual extinction. Figure 7.1 shows bass growth in an experiment with golden shiners.

When large gizzard shad (mean weight 454 g) were released in a pond at a rate of 68 per ac (168 per hec) in early winter (November) they required about five months before they were able to reproduce in the spring. A large hatch of young gizzard shad occurred on 30 April and although there were additional spawnings until late summer (August) these were much less important numerically. Bass had been released in the pond soon after the first shad had hatched, at a density of 200 per ac (494 per hec). The bass consumed these young shad and those from subsequent hatchings during the rest of the summer. Bass grew rapidly until they averaged 300·5 g in September, by which time they had eaten most of the small shad.

Because the shad spawning season was over little food remained for the bass. The food shortage was rendered more severe because the remainder of the shad, with competition for food reduced as a consequence of their own greatly lowered population density, grew rapidly in the late summer to become too large for bass to eat. By November shad young-of-the-year averaged 105 g and by that time the bass had actually lost an average of about 10 g each. Figure 7.2

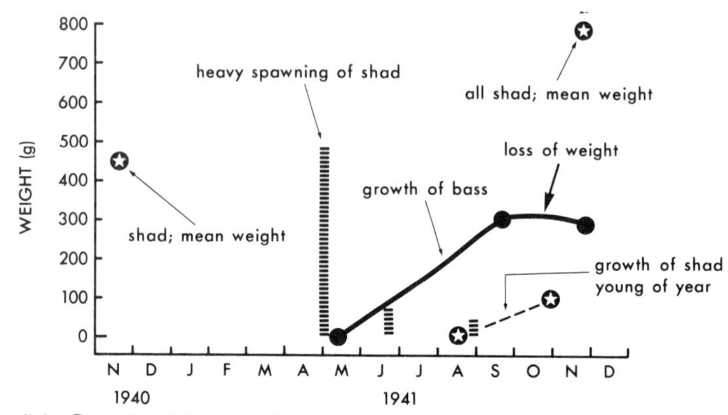

Fig. 7.2. Growth of largemouth bass as predators in fish pond in combination with gizzard shad as prey. See text for further explanation. (Constructed from the data of Swingle, 1946.)

depicts this course of events. When shad were combined with bluegills as "forage" fish for bass, interspecific competition for food prevented the attainment of normal size by adults so that their potential for production of young was seriously diminished.

In Alabama, goldfish usually spawn in spring (March to April). Swingle found that because the early broods ate the subsequently deposited eggs or the larvae hatching from them, only one batch of young goldfish usually survived. When bass were added to ponds as fingerlings in May they consumed many young goldfish and grew rapidly as a consequence until they reached a weight of about 200 g. By that time goldfish numbers were so depleted that survivors were able to grow much more rapidly and thus became too large for bass to eat. By then the goldfish had, under pond conditions, also become too large for successful production of young of their own, because of their cannibalistic tendencies. Bass, weighing 196 g at the beginning of this sharp drop in goldfish abundance, had little to eat, so that their mean weight actually decreased.

In one experiment (Fig. 7.3) goldfish were stocked in November to spawn in spring. The bass were added the following October. By

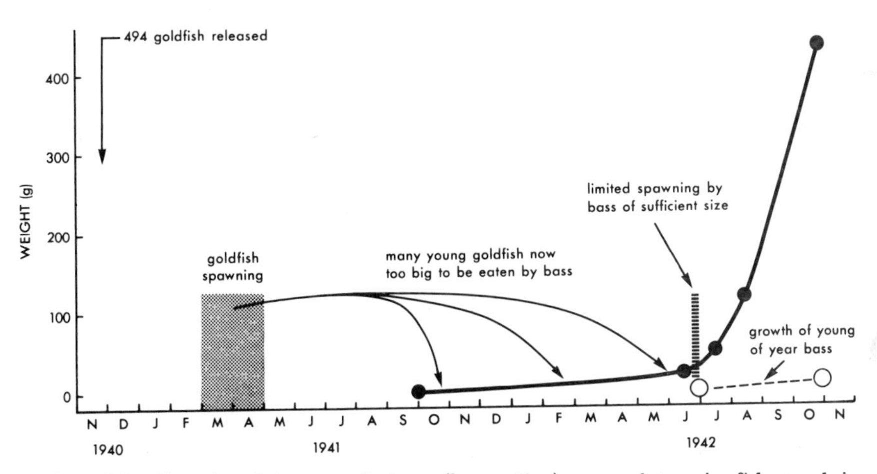

Fig. 7.3. Growth of largemouth bass (heavy line) as predators in fish pond in combination with goldfish as prey. See text for further explanation. (Constructed from the data of Swingle, 1946.)

that time many of the young-of-the-year goldfish had already become too large for the bass to prey on. Since the goldfish failed to spawn in their second year in the pond the fish of the first spawning became "so reduced (in numbers) that most of the survivors rapidly became too large for bass to eat." Actually, at the time the pond was drained there were remarkable numbers of forage fish—some 10 040 per ac (24 799 per hec)—too large, at an average of 26 g, to be preyed on. Swingle remarked that the goldfish would have been prevented from spawning in later years because of their own great numbers.

There may, then, be considerable actual complexity in the relationships between predators and prey even among apparently simple combinations of fish species. At least some aspects of these complex features would need to be allowed for in a fully rationalized culture of pond fish based on more than one species. It does not follow, of course, that predator-prey systems in nature closely resemble those in artificial fish ponds. But it is obvious that in certain kinds of situations predators must be profoundly influenced by the particular growth and reproductive characteristics of prey and that under otherwise similar conditions different species of prey fish may produce different predator responses. How rapidly prey grow in size determines, for instance, the length of the critical period during which they are vulnerable to predators. A corollary is that, if a long time elapses after a prey has been spawned before a potential predator contacts it, the effectiveness of the predator in coping with the prey may be seriously impaired. With this impairment comes slow growth, with its own several obvious concomitants, for the predator.

B. Some Principles of Predation

We turn now from Swingle's experiments to a hypothetical model as an aid to the closer examination of these problems.

Suppose there are two similar lakes, of which one receives water from a river in which a prey species of fish may spawn; the prey lives mostly in the lake. Suppose however that the young of the prey species normally spend the first 1½ years of life in the river before entering the lake. Suppose that the second lake, identical in most environmental respects to the first, lacks a river and has only small streamlets, big enough for spawning, but not for the sojourn of the young of the prey species which must accordingly appear immediately after hatching in the body of the lake. To both prey populations we assign 95% egg-larval mortality and assume identical predator popula ions in each lake, which do not become interested in the prey species until the latter are six months old. The essential differences between the lakes and populations are given in Fig. 7.4 and Table 7.1. Given the differences in behaviour of the prey

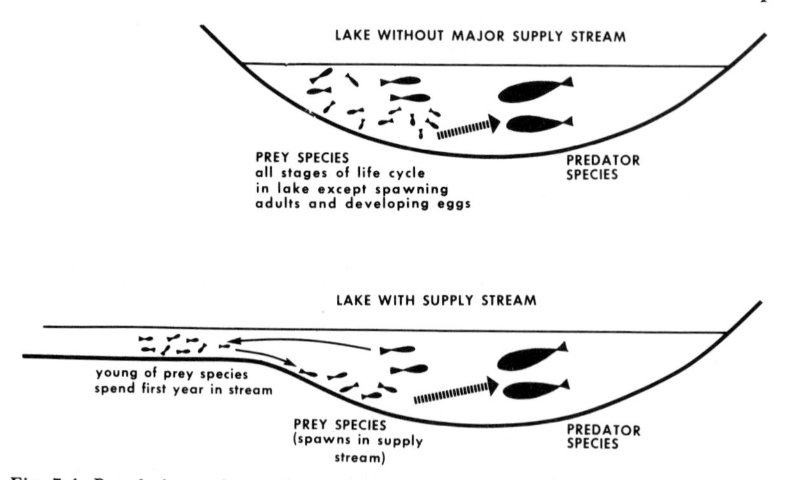

Fig. 7.4. Populations of a predator species and a prey species in lakes with and without major supply streams. In the former the prey spend entire life (except for spawning and egg-development) in lake. In latter young of prey are in stream for first year of life. See text and Table 7.1 for further information. (Original.)

Table 7.1

The effects on numbers surviving of a cohort of a prey population because of differences in the age at which predation first occurs

Age of Prey Cohort (yr)	Number Surviving: "Control" Cohort[a] (No predation)	River/Lake[b] Cohort (predation begins at age of 1½ yr)	Lake[c] Cohort (predation begins at age of ½ yr)
0[d]	500 000	500 000	500 000
1	250 000 (50)	250 000 (50)	200 000 (60)
2	125 000 (50)	100 000 (60)	80 000 (60)
3	62 500 (50)	40 000 (60)	32 000 (60)
4	31 250 (50)	18 000 (55)	14 400 (55)
5	15 625 (50)	9 000 (50)	7 200 (50)
6	7 813 (50)	4 500 (50)	3 600 (50)

The percentage mortality rates for the preceding 12 months are given in brackets after each cohort value.

[a] Annual mortality rate 50% of the preceding value.
[b] Annual mortality rate 50% except for the second, third and fourth years when it is increased by effects of predation.
[c] Annual mortality rate 50% except for first, second, third and fourth years when it is increased by effects of predation.
[d] All cohort sizes at age 0 are based on assumed 95% egg-larval mortality of an initial egg number of 10 000.

population imposed by the contrast in the environments, it can be stated that in the first lake the young fish are safe from predation until they enter the lake at 1½ years, whereas in the second lake they are preyed on one year earlier in their lives.

This assumes that, if there was no predation, post-larval mortality of the prey species would be 50% per annum. Table 7.1 shows the mortality we add to this each year of life as a result of predation. It is assumed that the proportion of fish eaten decreases with age and that prey older than four years are too large to be ingested. Table 7.1 shows the effects of predation on the survivorship in the two lakes. The earlier access of the predator to the prey, when the latter's whole life cycle (except for spawning itself) is within the lake, leads to a pronounced difference in the numbers of older prey surviving in comparison to the situation in which access to prey cannot occur during the first 1½ years of life. As there is frequently a close relationship between size of predators and size of prey (Popova, 1967) this could mean that age/size structures of the predator populations are closely tied to those of prey populations.

There is no need to pursue these questions exhaustively. The major principles exemplified in Swingle's work and further illustrated in this simple model provide us with sufficient departure points for analysis of various problems of this sort.

Swingle attempted to define the sorts of ratios between abundance of predator and prey that would lead to "balance" between populations. His approach was based on empirical experience rather than theoretical prescription (Swingle, 1950; Swingle and Smith, 1942).

Swingle (1950) correctly emphasized the formidable problems associated with adequate sampling of wild fish populations. His brief critical listing of the methods in use is still largely relevant. His own report was based on the fish populations in 89 well-established ponds from 2 to 30 years old in which the populations were censused by draining the ponds or in a few cases by killing the fish with rotenone. Swingle (1950) gave his most detailed prescriptions, based on his earlier work and developing theory, of the principles of population "balance". In stressing that this balance was delicately controlled he returned to his bluegill-bass example:

> Where bluegills are stocked in a fertilized pond at the rate of 1,500 per acre, approximately 1,000 will survive to reach a harvestable size of 4 ounces within 1 year and will completely utilise the food available for large bluegills. Each pair of bluegills then produces an average of 5,000 young during the first summer. In successive years, 4-ounce bluegills can be caught only if the young fish are reduced by predation and natural deaths to approximately the number removed by fishing plus the number of old fish that die annually. In that case, if both the adult parents are lost from these combined causes, there is room in the pond for two replacements to grow to a 4-ounce size; the predator fish and the natural mortality must, therefore, remove 4,998 and leave two replacements in this family group. Actually, for various species in various aquatic environments and for different rates of harvest, from 1 to 5 or more years will be required for the reduction in numbers of a brood to the correct number of replacements lost from these combined causes. This gives rise to the various "year-classes" and "length-frequency" groups below the size of the harvestable fish in a population.

He added (Swingle, 1950) that populations sometimes become "unbalanced"—incapable of producing numerous fish of satisfactory angling size yearly—through being unable to produce enough replacements for those dying to make satisfactory use of the available food supply. More frequently it was because survival of excessive numbers of fish made growth to a large size impossible because of intensity of competition for food.

An analysis of the results of Swingle's (1950) work will not be given. The basic concepts of predator-prey systems that he developed have already been described in his earlier experiments. This later work was an attempt to define various values of the predator-prey ratio in ponds that produced good fishing. As such, the results depend to some degree on the productivity of the ponds and the combinations of species used.

The whole point of Swingle's work for this chapter is, however, that it showed how importantly the growth pattern of one species could influence its own reproductive performance and the growth and reproduction of a predator species and that the reverse argument also held. Thus the biotic significance of these relationships stands revealed in the life of fish—and growth is the essential key to their evaluation.

C. Pike as Predators in Lakes

Gammon and Hasler (1965) and their colleagues Schmitz and Hetfeld (1965) have described a nine-year experiment on the effects of introducing muskellunge (pike) into two small lakes in Wisconsin. One, George Lake (17·3 hec), contained populations of yellow perch, largemouth and smallmouth bass. The other, Corinne Lake (14·6 hec), contained only yellow perch and largemouth bass. Both perch populations initially contained numerous "stunted" individuals.

In 1956, 400 yearling muskellunge *Esox masquinongy* were put in George Lake in May and, in October, 395 large young-of-the-year muskellunge were put in Corinne Lake.

Before 1956 the largemouth bass populations were thought to be fairly stable and numerically abundant, determined both by recapture estimates and fishing. The evidence for this appears more convincing for Corinne Lake. Gammon and Hasler observed good hatching of bass in both lakes, between 1956 and 1960, and attributed the decrease in numbers mainly to predatory inroads by muskellunge on the young bass which, after muskellunge were introduced, did not survive in sufficient numbers for satisfactory recruitment. The decline is reflected in the falling angling catch in which, after 1956, size increased while numbers decreased. However, by 1959, 1960 and later, small fish were again appearing in the catch in both lakes.

In George Lake, smallmouth bass showed neither of the same

tendencies for decreasing numbers and increasing size in the fished population. There was a dramatic change in the perch population of George Lake but it was already decreasing before muskellunge were released; some 19 000 were netted in 1963. After 1956 numbers greatly declined and, by 1957, insufficient perch could be captured to permit a valid population estimate. The perch of Corinnc Lake appear to have been stable at a high population density before 1956 and did not decline in numbers until 1959. However, by 1960 no estimate was possible, as was the case in George Lake by 1957.

Though perch were apparently abundant in both lakes before muskellunge were released, they were especially numerous in George Lake in 1955 and seem to have provided the major source of food for muskellunge which grew rapidly so that by the end of the first growing season they were already big enough to eat all but the largest perch. It appears that predation on young fish combined with the natural mortality of older fish caused the almost complete population collapse of perch already mentioned. Apparently the young muskellunge, when they were first released, lived largely on young-of-the-year perch, there being few small perch above that size. This was inferred from the abrupt deterioration of growth rate in muskellunge in George Lake after two seasons and after three seasons in Corinne Lake (Fig. 7.5). The growth of perch of the I and II groups increased considerably following the very severe reduction in numbers in Corinne Lake in 1960. In recent years, as muskellunge populations have gradually declined (20-25% annual mortality), the perch population has apparently begun to build up in the younger age groups.

The essential pattern of the effects of the muskellunge has been much the same on both largemouth bass and yellow perch. Their lesser effect on smallmouth bass was explained by Gammon and Hasler (1965) mainly in terms of behavioural differences, the distribution and more solitary habits of bass making them less liable to predation.

This was a study which consisted of a preliminary appraisal of some populations that were not changing much and at least one that was changing (or being changed) considerably. Into these populations was introduced a predator of a peculiarly effective kind. Even after a nine-year study it was impossible to say what the final outcome would be and the question was complicated by the gradual disappearance of the predator through mortality. However, as in the experiments by Swingle, it can be said that predators may show spectacular ability to change the numbers, year-class strengths and

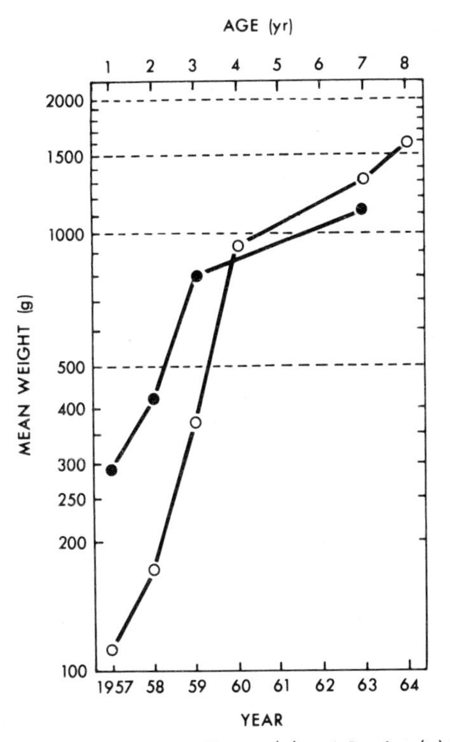

Fig. 7.5. The growth of muskellunge in George (●) and Corrine (○) Lakes. (After Schmitz and Hetfeld, 1965.)

growth rates of other fish. Thus it can be concluded that predators must considerably modify and control populations of other fish through their activities.

Munro (1957) reported that in a lake in Scotland in which other fish were not to be had as food, pike from 20 cm long began to consume other (smaller) pike. As the pike increased in size to 60 cm or more other pike came to assume growing importance (present in upwards of 40% of stomachs) in the diet. The only other available vertebrates—frogs—also became very important as food. The significant inference about this almost unavoidably piscivorous habit of large pike appears to be that, by cannibalism, pike can achieve a size and growth comparable to that in waters in which they have access to other fish species as prey.

Nikolskii (1969), in citing several examples of cannibalism among fish, has added some valuable observations to the concept of its functional significance in the population. He points out that perch are another well-known freshwater species which freely utilize their own young as food if other fish are absent and there is little benthic food available, because though the fry eat plankton the larger fish do not. He also notes that the young of salmon may feed directly on the corpses of their parents. A state of cannibalism may, according to Nikolskii, also develop in situations in which a particularly abundant year-class causes a food shortage for others. Nikolskii adopts the view that cannibalism can and does function as an adaptive mechanism to help populations exploit their food fully and therefore grow at optimum rates.

II. OTHER ASPECTS OF PREDATOR-PREY SYSTEMS IN FISH

In the most fundamental biological sense man's fishing activities must be regarded as a form of predation. But the nature of this predatory activity differs from that of the majority of natural predators. Fishing operations—except in those cases in which entire populations of fish are removed, as in pond culture or mass poisoning of a lake—tend to ignore small and young fish, concentrating only on individuals above a certain size, which may be determined by an agreed minimum or, automatically, by gear selectivity. In extreme cases fishing may be so effective as to eliminate virtually all fish above this critical size while leaving the remainder of the population intact. On the other hand, natural predation, whether by bird, mammal or other fish, may affect all age/size groups but will usually destroy many more small or young fish relatively than larger or older ones. Figure 7.6 shows an actual example based on Caspian roach, from Popova (1967) (data derived from Fortunatova, 1961); Nikolskii (1969) gives additional examples.

Predation may affect growth and competition not only among exploited age groups but also among other age groups that share the same food.

It is fairly remarkable that few additional specific examples can be given of predator-prey systems among fish. Beverton and Holt (1957) reviewed studies of fish preying on invertebrates but the far more interesting and—from our standpoint—vital question of fish preying on other fish remains largely unexplored. There is plenty of information about predatory fish which eat others, about the quantities they consume and under what conditions they do so. But the dynamic relationship between predator and prey, the estimation of the extent to which the life cycle characteristics of one species can be affected by another, remain largely unexplored.

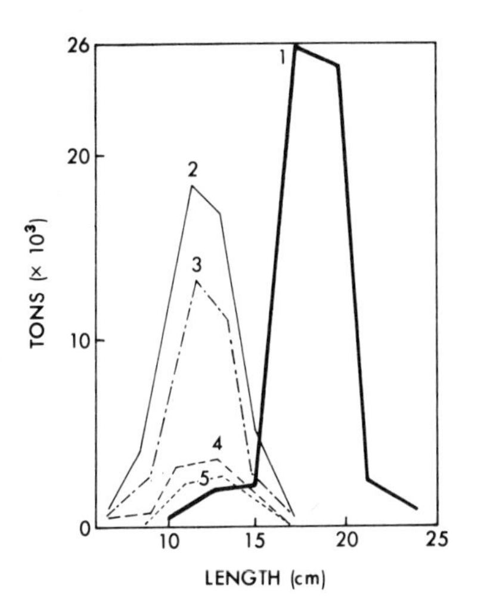

Fig. 7.6. Size composition of Caspian roach (vobla, *Rutilis rutilus caspicus*) removed from the waterbody (Fortunatova, 1961). 1, in catches; 2, in diet of all predators; 3, in diet of zander; 4, in diet of pike; 5, in diet of sheat-fish. (After Popova, 1967.)

Popova (1967), in a review of Russian literature on predator-prey relationships in fish, has presented some illuminating points. Russian workers have developed a method of potentially high significance for calculating the daily (24 h) ration of predators by a judicious mixture of field and laboratory observations. The ration is given by

$$X = \frac{(S_1 : n_1) + (S_2 : n_2) + \ldots + (S_n : N_n)}{V . N} . n$$

where

X = daily ration
$S_1 \ldots S_n$ = the amount of food eaten on different days, as a percentage of the weight of the predator
$n_1 \ldots N_n$ = the number of fish that fed on the day (24 h) in question
V = digestion rate in days
n = the number of specimens in the sample that contained food
N = the total number of specimens in the sample

Popova has explained that:

. . . experimental data provide estimates of digestion rate and the manner in which the food bolus disappears. In calculating the rations of predacious fish, reconstructed rather than actual weights of food components are used, since large prey objects undergo digestion over a rather long time, at unequal rates and often incompletely. Reconstruction of the live weight of organisms is done on the basis of their length, or from individual fragments of the prey: pharyngeal bones, lower jaws, otoliths, etc.

By the use of such procedures and the equation given, it has proved possible to compute daily, monthly and even annual rations for several species of commercially important fish.

Popova (1967), Smith, S. H. (1968) and Nikolskii (1969) stress the dynamic and opportunistic nature of the predator-prey relationship among fish. This is well shown in Smith's schematic dissection of the major predator-prey shifts that have occurred in the feeding relationships of deepwater species in Lake Michigan. Beginning before commercial exploitation, the records reveal the broad character of these shifts from 1910 until the present. Some of these relationships have values assigned to them and although they do not in any exact manner quantitatively signify the dynamics involved being based only on fishing or mortality statistics, it is obvious (Fig. 7.7) that changes in fishing pattern, the great effects of sea lampreys as predators since their entry in the 1930's and the introduction of other exotic species have combined in various complex ways, not fully understood, to produce the major recorded changes through the growth responses of populations.

III. PREY, PREDATORS AND GROWTH

Apart from such effects of predation on growth as described for fish populations in ponds, we can postulate, more generally, that there are many ways in which predator-prey relationships may affect growth or be affected by it.

For instance, many of the more strongly carnivorous fish are almost bound to consume other fish in order to grow beyond a certain size. Unless they can prey successfully, their particular functional morphology will not permit them to capture a sufficient quantity of food to grow further (Munro, 1957; Allen, 1935; Deelder, 1951).

Conversely, it could be postulated that many examples of natural fish populations exist in which abundant reproductive success will lead to such heavy competition for food that growth is virtually absent. Predation could here be invoked as a process tending to regulate some such populations so that they do not become overcrowded.

Among animals whose growth is competition-dependent, such as

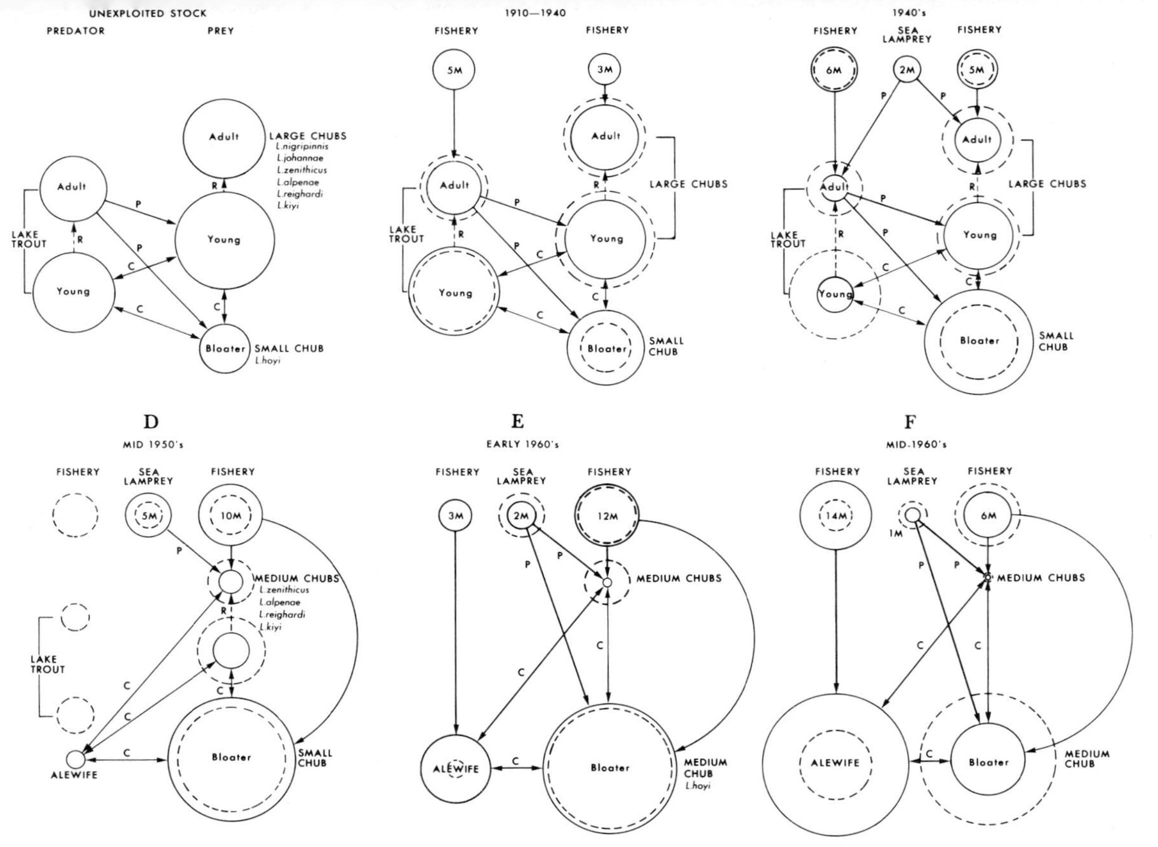

Fig. 7.7. Interrelations of major deepwater species of Lake Michigan before exploitation (A) and during the following periods: stable exploitation, 1910-40 (B); early influence of the sea lamprey, 1940's (C); maximum abundance of the sea lamprey, mid-1950's (D); maximum abundance of bloaters, early 1960's (E); maximum abundance of the alewife, mid-1960's (F). P = predation; C = food competition; and R = recruitment. (After Smith, S. H., 1968.)

fish, it appears quite likely that predator-prey systems help to ensure a natural balance both within and between populations. Such systems could be expected to function, via effects on growth and thence reproduction, in such a manner as to impose a stability on populations of both predator and prey.

8 | Production

Lindeman (1942) and Borutsky (1939a, b) were among the first ecologists to give serious and detailed consideration to the identification and measurement of production processes in natural ecological systems. Since then, many valuable studies have been reported (e.g. Odum, H. T., 1956, 1957a, b; Odum and Odum, 1955; Odum and Hoskin, 1957, 1958; Beyers and Odum, 1959). The subject has also found its way into ecological textbooks as a standard, indeed unavoidable, topic. Recently the entire question of production in fish has been extensively and penetratingly reviewed (e.g. Gerking, 1967; Ricker, 1968).

Production processes are usually analysed with reference to the rates at which matter moves between the trophic levels of ecosystems. Plants and other autotrophs, using solar energy, impose a relatively high degree of biochemical organization on such simple substances as carbon, nitrogen, oxygen and hydrogen—together with sulphur, calcium, magnesium, sodium, potassium, phosphorus, chlorine and small quantities of many other elements. The autotrophs impose organization by incorporating these simple substances into organic molecules—carbohydrates, fats, proteins and other kindred materials of greater or lesser complexity—of which living tissues are composed. These synthesized organic materials comprise the sole primary source of food substances available to animals and represent not only the most chemically complicated and diverse range of compounds known, but also a state of matter high in "stored" energy, which is therefore thermodynamically "improbable".

As they are transferred along food chains organic molecules tend to become structurally degraded, as when a carbohydrate is oxidized, or to become organized into still more complex substances, as when amino acids combine to form proteins. The organization of these molecules within the body may occur at two different levels: that concerned with the substances they unite to form, and that concerned with the tissues and organs made of the molecules and substances. At certain times it has been suggested that such organizations of matter within organisms—and therefore the existence of organisms themselves—might represent exceptions to the second law of thermodynamics. In fact, however, it appears that their synthesis, maintenance and degradation are all closely governed by the use of precisely determined amounts of energy. In the case of animals, such energy can be derived only from the oxidation or fermentation of already existing organic materials.

The only significant basic energy source available to any ecosystem is solar and because every link in a food chain and step in a metabolic pathway involves energy expenditure, the energy in an ecosystem has a flux—but not a cycle (Odum, E. P., 1959; Macfadyen, 1963).

Like physiologists, ecologists have long realized that the mere size of an organism, whether or not it is growing, signifies little about the rate at which it incorporates matter into its body, utilizes it or excretes it. It is also well-known that, when speaking of organic matter and the energy it contains, one must necessarily specify the particular form of the organic matter and its role in the body's economy. Thus, fat contains a relatively large amount of energy but a mass of fat, stored in the body's reserves or present as an insulating layer, may in itself be chemically and biologically inert. For eventual liberation of its energy as heat to occur, fat has to become incorporated into some biochemically functional part of a living system.

It should be obvious, then, that production, denoting transposition and transformation of materials (implying growth and excretion etc.), can frequently be measured, or at least quantitatively rated, in terms of its accompanying energy demands, incorporations and releases.

Many of the ideas underlying this sketch of production have already been given (Chapters 2 and 4). They are re-presented here in this simple and more broadly generalized manner in order to establish certain basic points of view before turning to the quantitative analysis of production in fish populations.

I. PRODUCTION IN THE POPULATION

Chapman (1967, 1968) has recently given an excellent general account of production in the fish population and much of what follows depends on his exposition. Chapman explained that he was dealing with that aspect of production (P) which refers to the total weight of living fish produced in a given time interval (Δt). This is essentially the concept of production already outlined. Chapman's statement of production in gravimetric terms is:

$$P = B_1 - B_0 + B_d$$

where

$B_1 - B_0$ = net biomass change

and B_d = total weight (at death) of all fish dying, during time Δt

Chapman also pointed out that

$$B_A = P + B_r$$

where

B_A = weight of food assimilated during Δt
B_r = weight loss by fish due to respiration during Δt

Therefore:

$$B_c = P + B_r + B_u + B_v = \text{consumption of food}$$

where

B_c, B_u and B_v = weight of food consumed, faeces and urine produced, respectively, during Δt

This accounting of production parallels that of Needham (1964), Kleiber, (1961), Davies (1967) and Warren and Davis (1967) as given in Chapters 2 and 4 where the values of various appropriate terms, which can be made as numerous as required, are estimated and the terms are then summed to give production or, by difference, various other values.

In computing production in populations of fish:

$$\frac{dB}{dt} = (G - Z)\,B = \text{rate of change of biomass of the stock}$$

where

G and Z = instantaneous growth and mortality rates, respectively, during Δt

More simply, according to Allen (1950) and Ricker (1946):

$$P = G\bar{B}$$

where

P = production,
G = instantaneous rate of increase of weight,
$\bar{B}$ = mean biomass during Δt.

Chapman (1968) considers this to be a realistic formula if G is known to be constant during Δt.

For annual production by a fish population Chapman developed a formula which, though based on exponential rates, did not include the assumptions that either individual fish or fish populations grow or die in an exponential manner. He also admitted that the mean of two estimates of biomass based on actual data was usually to be preferred to biomass estimates from models of exponential growth. However, he referred to Allen's (1950) advice that field work can be reduced by employing a suitable expression to account for growth over long periods. Two formulas (based on Brody, 1927) are suggested as describing sigmoidal growth in animals:

$$Wt = Wo\mathrm{e}^{gt} \tag{8.1}$$

where g is an instantaneous rate of change and

$$Wt = A - B\mathrm{e}^{-gt} \tag{8.2}$$

where A and B are constants.

Formula (8.1) describes early exponential growth either in individuals or in numbers of a population. Formula (8.2) accounts for the decaying, post-inflection part of the curve as it approaches an upper asymptote. There is an obvious parallel between these formulas and those discussed in Chapter 4, though they have somewhat different derivations. It appears from experimental observations with very young fish that growth in weight and increase in numbers are both best accounted for by exponential functions.

This formula assumes that the population is in steady state—i.e. has constant annual recruitment and mortality throughout the life of the fish. For older fish, however, a Bertalanffy model may provide a better description of growth.

Beverton and Holt (1957) proposed a formula for production (P) based on the Bertalanffy growth model:

$$P = 3K\,(W\infty^{1/3}\ \bar{B}W^{2/3} - \bar{B})$$

where

K = one of two main parameters of the Bertalanffy equation

$W\infty^{1/3}$ = the asymptote of the curve of weight growth to the 1/3 power

$\bar{B}$ = mean population biomass for the year

$\bar{B}W^{2/3} = \int_0^1 Wt^{2/3} Nt^{dt}$, which is in part proportional to the mean surface area of the fish in the population during the year.

Allen (1951) originated a graphical means of determining production in which progressive values for number of fish are plotted against mean weight of individuals in a population cohort as these values change with time. The area beneath such a curve represents production which can therefore be determined for any part of the year. Allen computed production in a trout population from curves of this sort. He had to assume that growth and survivorship patterns were similar in different sections of the stream and that they recurred fairly constantly each year—assumptions which were apparently justified in the case he considered. More general applications of this graphical method would, of course, demand caution but its strength lies in the fact that the shape of the curve depends on actual data on survivorship. When these, together with useful estimates of mean weight of fish, are available adequate quantified assessments of production should be obtainable.

The mathematical equivalent of determining production, by measuring the area under an Allen curve by means of a planimeter, is shown in Fig. 8.1. Here, during a small period of growth Δw at time t, ΔWNt is the quantity of new material produced. Integration in the function for production between times t_0 and t_1 gives

$$P = \int_{w0}^{w1} Nt\,d\bar{W}$$

Allen's method may of course be applied only *a posteriori* and fragmentary data are inadequate for computation of production. However, as the number of satisfactory estimates (see Chapman, 1967) of fish production is small, it seems desirable to employ a method which can lead to accurate production assessments even if it does also require the collection of numerous data.

Chapman (1967, 1968) usefully listed some important factors that may distort assessments of mean growth of individuals in a population or a cohort; among which are mortality rates that differ as between individuals of different size groups, loss or larger

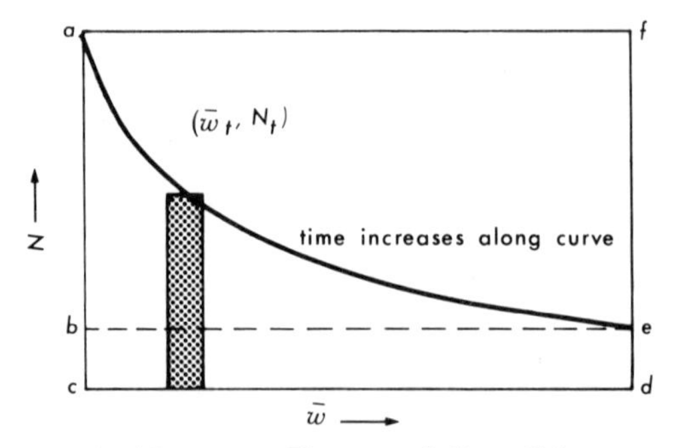

Fig. 8.1. A schematic Allen curve. The sum of all small increments $\bar{w}Nt$ is production between times *a* and *e*. Shaded area is such an increment. (After Chapman, 1967.)

individuals through migration and seasonal degrowth and weight loss through release of ova or sperm. Though these and other causes of confusion can often be readily allowed for in principle they may often be difficult to quantify.

Chapman (1968) has also given an extended, rigorous treatment of the full theoretical methodology for computing fish production, illustrating his account with some practical examples.

We now turn to two major source studies of particular interest.

A. Cultus Lake Salmon

Ricker and Foerster (1948) were responsible for a notable early investigation of production of the sockeye salmon of Cultus Lake in British Columbia, which offered exceptional advantages for such work, because the entire early life of the salmon is passed within the lake and the stocks could be censused at the time of the seasonal migration seawards. In Cultus Lake, sockeye spawn from October to December and the young hatch in spring; mid-May was considered the beginning of free-swimming life, for convenience of computation. Young fish quit the lake during the following April and May, 30 April being taken as a representative departure date. A variable proportion of O group fish remain in the lake for a second year but such fish (I group) were usually considerably less numerous than the O group yearlings leaving the year before. A third group of sockeye was found to remain permanently in the lake. Though variable in number they comprised but a minor fraction of the total lake population, and were therefore not counted.

Because of the particular nature of the population age structure of Cultus Lake sockeye, resulting largely from the migratory behaviour of the fish, Ricker and Foerster proposed that:

$$\text{Annual sockeye production} = \begin{array}{c}\text{total weight}\\\text{increase of}\\\text{O group during}\\\text{year}\end{array} + \begin{array}{c}\text{total weight}\\\text{increase within}\\\text{I group during}\\\text{year}\end{array}$$

Naturally, this total was obtained only by summing the weight increments of those fish surviving and dying during the year.

Direct counts of migrants and knowledge of their mean weight were available during all the years of the study (Foerster, 1944), but number and weights of non-migrants (i.e. O group remaining) were less certain. Average weight of non-migrants was, however, about 50% less than that of migrants at the time the latter left the lake.

The mean weights of young sockeye were obtained during the year from their presence in the stomach contents of predatory fish of other species. The weights were calculated from lengths, it being impossible to weigh satisfactorily most of the ingested sockeye, as they were already partly digested.

The conversion formula used was

$$W = 0{\cdot}0100L^3$$

The authors explained that though this relationship did not hold well for very young sockeye, the origin of the curve of weight increase could be satisfactorily "anchored" at 0·15 g, the mean weight of the fry at the time they became free-swimming.

Ricker and Foerster concluded that the curves of mean growth of individuals of the O group of three separate years resembled one another sufficiently (Fig. 8.2) to suggest that the general pattern of weight increase was of the same order from year to year. Consequently, the 1931 year-class was taken as typifying growth of young sockeye in Cultus Lake. (The authors admitted that the slightly faster growth of 1933 could be inversely correlated with smaller numbers in that year; see Table 8.1.)

Instantaneous growth rates for each half-month were then determined from the growth curve for the 1931 year-class by reading logarithms of weights from the curve, calculating the differences between successive values and dividing by 0·4343. The instantaneous growth rate for 1931 as a whole—the sum of all the semi-monthly rates—was 3·01. Ricker and Foerster now applied instantaneous growth rates for each half-month proportionately to all other years for which they possessed growth data and for all year-classes. An example, for the 1934 year-class, was computed as follows:

mean weight of yearling stock = 8·79 g
mean weight of fry = 0·15 g

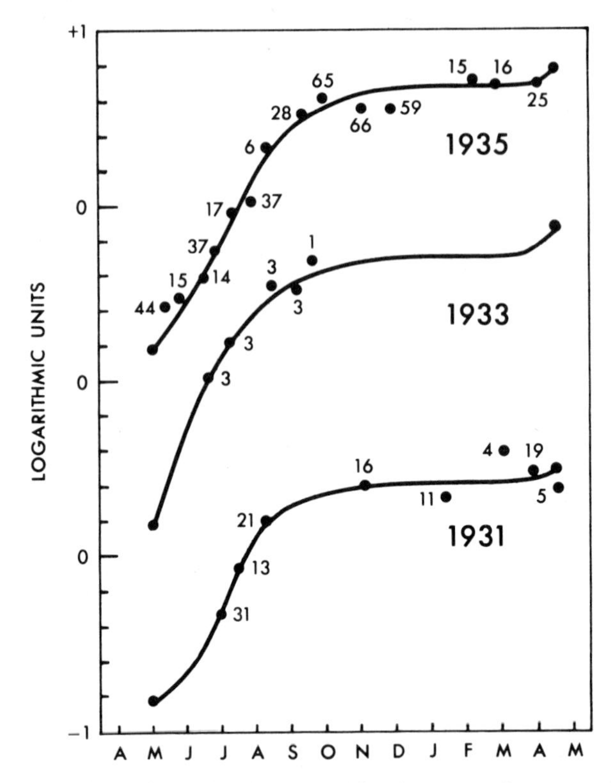

Fig. 8.2. Logarithms of calculated mean weights in grams of young sockeye salmon, taken from stomachs of predators, of the year-classes of 1931, 1933 and 1935 in Cultus Lake. Number beside each point indicates the number of fish which it represents. The first point in each series is the mean weight of free-swimming fry; the last point is mean weight of the yearling stock. (Table 8.1, column 10.) In the ordinate, each zero applies to the curve it stands opposite to. (After Ricker and Foerster, 1948.)

(i.e. yearlings are 58·6 times heavier than fry). Instantaneous growth rate for year = $K = \log_e 58{\cdot}6 = 4{\cdot}07$.

Thus, in determining semi-monthly instantaneous growth rates for 1934 fish, each semi-monthly value for 1931 fish was multiplied by 4·07/3·01 (assuming the same general pattern of growth).

Computation of the semi-monthly mortality rates was based on survival rates for marked fish of the 1933 year-class, released at various times of the year. These led to the mortality rates in Table 8.2. The sum of these semi-monthly mortality rates equals the annual instantaneous mortality rate ($i = 2{\cdot}80$). This represents the mortality before predator control, which operated between 1934-36 inclusive and consisted of selective removal of fish of other species that consume young sockeye.

Table 8.1

(After Ricker and Foerster, 1948)

Number and weight of year-old migrants, nonmigrants and total yearling stock

1	2	3	4	5	6	7	8	9	10
	Migrants			Nonmigrants			Total Yearling Stock		
Year-class	Number (thousands)	Average Weight (g)	Total Weight (kg)	Number (thousands)	Average Weight (g)	Total Weight (kg)	Number (thousands)	Average Weight (g)	Total Weight (kg)
1925	183	8·10	1 480	17	7·0	120	200	8·00	1 600
1926	336	5·04	1 690	83	3·5	290	419	4·73	1 980
1927	2426	3·06	7 420	666	1·68	1120	3092	2·76	8 540
1928	39	6·55	260	52	3·0	160	91	4·62	420
1929	350	7·10	2 490	2	7·0	10	352	7·10	2 500
1930	788	7·32	5 770	0	—	0	788	7·32	5 770
1931	1571	3·67	5 770	633	1·46	920	2204	3·04	6 690
1932	121	6·53	790	142	4·0	570	263	5·17	1 360
1933	242	7·55	1 830	4	7·5	30	246	7·55	1 860
1934	502	8·83	4 440	69	8·6	590	571	8·79	5 030
1935	3101	5·96	18 490	60	5·8	350	3161	5·96	18 840
1936	1627	7·2	11 720	10	7·0	70	1637	7·2	11 790

Table 8.2
(After Ricker and Foerster, 1948)

Computation of relative biomass, production and mortality in successive half-months from the data of Table II of Ricker and Foerster.

Column	2	3	4	5	6	7	8	9	10	11	12
	Computation of biomass, production and mortality for the year-class of 1931								Biomass, production and mortality for the year-class of 1935		
Date	k	i	k–i	Change in Biomass	Biomass	Average Biomass	Production	Mortality	Biomass	Production	Mortality
May 1					100				100		
	0·07	0·27	–0·20	–0·18		110	8	30		10	20
Apr. 16					122				111		
	0·05	0·10	–0·05	–0·05		125	6	12		7	8
Apr. 1					128				112		
	0·02	0·04	–0·02	–0·02		130	3	5		2	3
Mar. 16					131				113		
	0·00	0·27	–0·27	–0·24		151	0	41		0	24
Dec. 16					172				137		
	0·06	0·17	–0·11	–0·10		181	11	31		10	17
Nov. 1					192				144		
	0·05	0·06	–0·01	–0·01		192	10	12		9	6

Oct. 16					193				141		
	0·09	0·07	+0·02	+0·02		191	17	14		15	7
Oct. 1					190				133		
	0·09	0·09	0·00	0·00		190	17	17		14	8
Sept. 16					190				127		
	0·14	0·09	+0·05	+0·05		185	26	17		21	7
Sept. 1					181				112		
	0·28	0·11	+0·17	+0·19		165	47	18		33	8
Aug. 16					152				86		
	0·41	0·12	+0·29	+0·34		131	54	16		35	6
Aug. 1					113				57		
	0·60	0·15	+0·45	+0·57		91	55	14		31	4
July 16					72				30		
	0·49	0·20	+0·29	+0·34		62	31	13		15	3
July 1					54				19		
	0·35	0·24	+0·11	+0·12		51	18	12		7	3
June 16					48				15		
	0·19	0·31	–0·12	–0·11		51	10	16		3	3
June 1					54				14		
	0·12	0·51	–0·39	–0·32		66	8	34		2	6
May 16					80				17		
Totals	3·01	2·80	+0·21	+0·23	81		322	302	17	216	133

Ricker and Foerster compared survival of sockeye fry liberated before predator control with that afterwards, calculating geometric means for survival of 3·63 and 10·20%, which corresponded to instantaneous mortality rates of 3·32 and 2·29 respectively. On the basis of this ratio and given $i = 2{\cdot}80$ for the earlier period, $i = 1{\cdot}93$ for the later period of predator control. This, in turn, corresponded to a survival rate of 14·5%, only a little greater than the observed 13·1% survival of fry.

Ricker and Foerster observed that "the production of fish population is the sum of the growth made by all its members. . . ." If growth and mortality can be assumed not to change during the year then:

$$\text{average population} = \frac{W_0\,(e^{k-i}-1)}{k-i} \tag{8.3}$$

where

W_0 = initial weight of stock
k and i = instantaneous growth and mortality rates, respectively, for the year

Because, however, in Cultus Lake sockeye both k and i were subject to sharp seasonal changes, separate computations of average population for short time intervals had to be made.

In computing production Ricker and Foerster first determined differences between instantaneous rates of growth and mortality† during the year. Biomass change was then calculated from an exponential table, using the value $(k-i)$. By then assuming a value of 100 weight units for the yearling stock at the time of migration they computed biomass at preceding semi-monthly intervals for the whole previous year (Table 8.2). Negative values of $k-i$ indicate that biomass decreased; e.g. on 15 April its size was calculated as $100/(1-0{\cdot}18) = 122$ weight units. The average biomass during the same period is determined from (8.3) and is 110. All the other biomass values in Table 8.2 were compiled in a similar manner.

Production during each semi-monthly period is the product of k times the average population biomass. This value (column 8 in Table 8.2) could alternatively be expressed as the area under an Allen curve for each semi-monthly period. Similarly, weight of fish dying in a semi-monthly period is the product of i times the average

† In the sense of Ricker and Foerster (1948) "mortality" signifies the *weight* of fish dying. Thus $k-i$ represents the difference between the gain of weight due to mean instantaneous growth rate of individuals in the population and the continual weight loss by the population through mortality.

Table 8.3
(After Ricker and Foerster, 1948)

Production of, and mortality among, age 0 sockeye during their year of lake life (in kg)

1 Year-class	2 Weight of Yearling Stock	3 Initial Weight of Fry	4 Mortal-ity	5 Produc-tion	6 Ratio of Production to Yearling Stock	7 Ratio of Production to Mortality	8 Summer Production
1925	1 600	500	3 750	4 850	3·03	1·29	2 960
1926	1 980	1030	5 280	6 230	3·15	1·18	4 140
1927	8 540	7500	26 500	27 500	3·22	1·04	19 300
1928	420	220	1 130	1 320	3·15	1·17	880
1929	2 500	870	7 620	5 020	3·05	1·29	4 720
1930	5 770	1900	13 700	17 600	3·06	1·29	11 000
1931	6 690	5300	20 200	21 500	3·22	1·06	14 900
1932	1 360	650	3 520	4 270	3·14	1·21	2 760
1933	1 860	610	4 390	5 700	3·05	1·30	3 300
1934	5 030	590	6 140	10 700	2·13	1·74	6 000
1935	18 840	3300	25 100	40 700	2·16	1·62	23 700
1936	11 790	1700	15 100	25 200	2·14	1·67	14 500

biomass of the population. To distribute production within a real population for the months of any particular year Ricker and Foerster needed to know only the actual biomass of the yearling stock at the time of migration into the sea. In the present case total annual production is 322, calculated from 100 weight units of biomass at migration. The *actual* biomass at that time (Table 8.3) was 6690 kg, so that:

$$\text{production} = 322 \times 6690/100 = 21500 \text{ kg}$$

and

$$\text{mortality} = 302 \times 6690/100 = 20200 \text{ kg}$$

In these computations the logical steps are based on few factual data, a situation common to studies of this sort. However, as Ricker and Foerster pointed out, the results can be checked at two points. In the first place, the initial stock of 80 weight units, found by the successive divisions down column 6 (Table 8.2) can be checked by direct calculation from the total of the $k-i$ column. This equals +0·21, which corresponds to a change of +0·23. The initial stock therefore equals 100/1·23 = 81 units of weight, which is approximately equal to 80. The second check involves a comparison of gains and losses for the year. The weight of the fry at the beginning added to production should balance, or equal, the weight of the yearling fish plus the weight of those fish that died during the year. The sums are 80 + 322 = 100 + 302 = 402, which of course represents remarkably good agreement!

Figure 8.3 shows biomass, growth (mean weight against time), the decline in population, production and mortality plotted to show their trends over the course of a year (in this case for the 1933

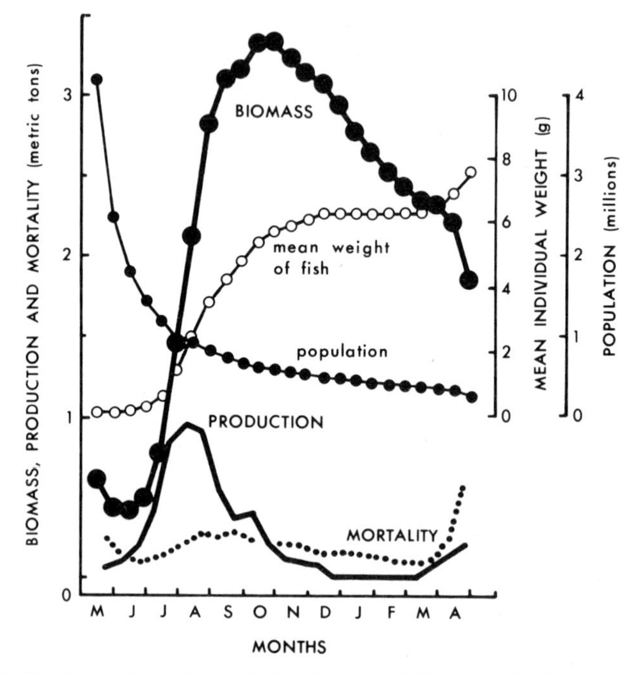

Fig. 8.3. The interrelationships of abundance, individual weight, biomass, production, mortality, for the O group sockeye of the 1933 year-class in Cultus Lake. Note that the difference between production and mortality for any half-month equals the net increase (or decrease) in biomass during the same period. (After Ricker and Foerster, 1948.)

year-class). The major (2/3-3/4) of total annual production always occupied the summer months—the period mid-June to mid-September—in Cultus Lake. Mortality, though more evenly distributed, had a spring maximum and a winter minimum.

Ricker and Foerster (1948) used much the same procedures as those already described to determine production in I group sockeye. As there were usually many fewer fish than in O group, data on growth, mortality etc. of I group were much less adequate and are not considered here. There was also an attempt to relate fish production to food consumed, which was relatively unsuccessful because of untestable assumptions about turnover rates of the entomostracans that comprised the food.

B. Trout of the Horokiwi Stream

Allen's (1951) basic plan in his major study of the trout population of the Horokiwi Stream, near Wellington, New Zealand was "to attempt a comprehensive quantitative study of a small isolated trout population, the factors by which it was controlled, and, particularly, its relation to its food supply." The choice of the Horokiwi was based on a number of preconceived requirements for a stream suitable for investigation, which included suitable size (7-10 miles long), minimal branching, separate entry to the sea, full trout stock, ease of netting, accessibility and absence of pollution.

The 10-mile-long Horokiwi consists of two main branches of similar size, joining about three miles inland. Allen divided the stream into a number of zones characterized in terms of their contained water types (pools, runs etc.).

The main method for sampling trout was to secure stop nets at representative sampling sites and seine the enclosed stream thoroughly with small mesh nets until no more fish could be caught. Some of this netting was done at night by torchlight, this being especially effective in collecting fry which spend the day in the interstices between stones. It was thus also possible to compare the day and night feeding of trout. Some captured fish were killed for examination of stomach contents, others were placed in live-boxes for subsequent measurement and sometimes for tagging or marking before being returned to the stream.

Redds excavated by females were counted during the spawning season and their position noted. A sample of the redds was dug out and the eggs, collected in a fine drift net placed downstream of the excavation, were counted.

Angling records were also obtained from those fishing the Horokiwi.

(a) Trout Movement

The Horokiwi trout displayed only a very limited range of movements. Of 76 trout recovered after tagging only four were found outside the river zone in which they had been released and three of these four were in adjacent zones. Of all the fish recovered after tagging 80% were found less than 250 ft from their initial liberation points. The times that elapsed between tagging and recapture (which ranged from a few to 333 days) evidently bore no consistent relation to distance travelled from point of release and most Horokiwi trout apparently remained permanently in a rather small home range, a fact which allowed Allen to ignore the possibility that growth patterns of particular trout might be the composite results of growth in different sections of the stream.

(b) Estimation of Numbers

Allen (1951) reviewed the difficulties of obtaining population estimates from mark-and-recapture data, especially in stream populations such as the Horokiwi where the trout move so little. Allen treated every sampling region as if it were a self-contained and limited area and estimated numbers of adult trout from mark-and-recapture data. Eggs and fry were estimated by two methods: from egg counts on mature females and from excavated redds. Trout were hard to catch in the first few months after hatching, so data on their mortality and growth during that period were scanty.

Since the various zones and regions of the Horokiwi were regarded, in practice, as discrete units, Allen proposed to extrapolate population estimates based on these regions to the stream as a whole. He considered difficulties such as possible differences in mortality and growth rates between tagged and untagged trout and movement of marked fish out of the zones they were tagged in. On the first point, Allen's data from marked fish suggested that 20% left a particular stream section in the intervals between routine nettings. Loss of tags complicated the estimates and an attempt was made to measure rate of tag loss in the 1940 year-class by tagging their dorsal fins and clipping their adipose fins. Unfortunately the sample recovered was too tiny to permit valid inferences about rate of tag loss but other evidence suggested a rate of about 30% over three months.

Table 8.4 shows the recaptures, during the routine netting of the Horokiwi of 1940 and 1941, of previously tagged fish. The variation revealed on analysis was $\chi^2 = 6{\cdot}22$ with a probability of natural occurrence of 0·29 for five degrees of freedom. The netting efficiency was essentially the same for all stations, the mean value being 0·103 for the proportion of marked fish recovered. This factor was arbitrarily adjusted to 0·15 to compensate for the loss of tags between routine nettings and for emigration out of zones by about 20 and 30%, as previously mentioned. Allen tested statistically the validity of these procedures and concluded that "the uncorrected

Table 8.4

(After Allen, 1951)

The numbers of fish of each year-class caught and estimated to have been present in each zone at each routine netting

Year-class	Date	Number caught in zones I.	IIM.	IIR.	III.	IV.	V.	Total	Estimated total populations in zones I.	IIM.	IIR.	III.	IV	V.	Total
1940	Jan. 41	99	69	. .	11	21	12	212	6330	2250	500	1950	800	170	12 000
	May 41	178	247	50	49	19	18	561	2520	3175	725	700	240	150	7 500
	Oct. 41	2	27	. .	1	13	5	48	140	800	180	170	550	60	1 900
1939	July 40	12	62	. .	4	16	14	108	750	1875	425	675	600	175	4 500
	Oct. 40	. .	37	. .	3	25	16	81	650	1130	250	520	950	200	3 700
	Jan. 41	14	28	. .	2	13	10	67	900	920	200	350	490	140	3 000
	May 41	21	54	11	45	36	10	177	300	680	160	625	450	85	2 300
	Oct. 41	0	6	. .	0	4	7	17	0	175	40	0	170	90	475
1938	July 40	5	4	. .	3	6	1	19	300	190	40	425	250	15	1 220
	Oct. 40	. .	6	. .	3	7	0	16	310	200	40	450	270	15	1 285
	Jan. 41	2	2	. .	0	3	1	8	80	50	10	120	70	5	335
	May 41	2	4	0	3	3	1	13	40	25	5	60	35	2	167
	Oct. 41	0	2	. .	0	0	0	2	18	12	3	25	15	1	74
1937	July 40	1	4	. .	0	10	1	16	140	80	25	0	355	10	610
	Oct. 40	. .	0	. .	0	6	1	7	65	40	10	0	175	5	295
	Jan. 41	2	1	. .	0	2	0	5	55	30	10	0	145	5	245
	May 41	3	1	1	0	5	0	10	30	17	5	0	75	3	130
	Oct. 41	0	0	. .	0	0	0	0	0	0	0	0	0	0	0
1936	July 40	0	0	. .	1	1	1	3	10	0	10	130	60	10	220
	Oct. 40	. .	0	. .	0	1	0	1	2	0	2	24	10	2	40
	Jan. 41	0	0	. .	0	0	0	0	2	0	2	20	10	1	35
	May 41	1	0	1	0	0	0	2	2	0	2	18	7	1	30
	Oct. 41	0	0	. .	0	0	0	0	0	0	0	0	0	0	0
1935	July 40	0	1	. .	0	0	0	1	. .	. .	. .	. .	. .	. .	40
	Oct. 40	. .	0	. .	0	0	0	0	. .	. .	. .	. .	. .	. .	32
	Jan. 41	0	0	. .	0	0	0	0	. .	. .	. .	. .	. .	. .	12
	May 41	1	0	0	0	0	0	1	. .	. .	. .	. .	. .	. .	6
1934	July 40	1	0	. .	0	0	0	1	. .	. .	. .	. .	. .	. .	40
	Oct. 40	. .	0	. .	0	0	0	0	. .	. .	. .	. .	. .	. .	32
	Jan. 41	0	0	. .	0	0	0	0	. .	. .	. .	. .	. .	. .	12
	May 41	1	0	. .	0	1	0	2	. .	. .	. .	. .	. .	. .	6

(Aft

The numbers of fish of each year-class, caught in each station, in the ne

Zone	Netting Station	Equiv. Routine Netting Station	Number of Fish Taken Year-class 1940	1939	1938	1937	1936	1935	1934	T
I	MA		5	5	..	..	..	..	..	
	MB		30	2	..	..	..	..	..	
	MC	M II	42	4	..	..	..	..	..	
	MD		51	6	..	2	1	..	1	
	ME		50	4	2	1	..	1	..	
II M	MF		54	12	1	..	..	..	..	
	MG	M IV	34	..	1	..	..	..	..	
	MH		99	25	1	..	..	..	..	1
	MI	M V	60	17	1	1	..	..	..	
II R	RA		45	8	..	..	..	..	..	
	RB		5	3	..	1	1	..	..	
III	RC	R II	5	5	2	..	..	..	..	
	RD		20	11	..	..	..	..	..	
	RE		24	29	1	..	..	..	..	
IV	RF		6	22	2	3	..	..	..	
	RG	R IV	13	14	1	2	..	..	1	
V	RH	R V	18	10	1	..	..	..	..	
		TOTALS	561	177	13	10	2	1	2	7

estimate of netting efficiency can be stated with 95% confidence to lie within the range 0·07 to 0·14. The corresponding corrected values are 0·10 to 0·21." Armed with this factor of netting efficiency (0·15), Allen computed number of fish of each year-class caught at the various stations in an experimental period of netting (April to June 1941); see Table 8.5. He then extended this analysis to cover fish of each year-class caught and estimated to have been present in each zone at every routine sampling during 1940-41. In applying his factor of 0·15 Allen provided the following tabulation of the range of 95% confidence limits for samples of different sizes:

5 fish	+150 to −65%
20 fish	+50 to −35%
50 fish	+33 to −22%
100 fish	+22 to −15%

The relative error was therefore greatest for the older year-classes, for which it could range from +250% to −75% (for a sample of only three fish).

.5
, 1951)

pril to June, 1941, and the total numbers estimated to have been present

st. otal mber esent	Station	No. per 10^4 sq. ft. Zone	Total	Estimated Number in Zone Year-class 1940	1939	1938	1937	1936	1935	1934
56	40									
13	137									
07	239									
08	279									
36	295	189	2 942	2530	299	28	43	14	14	14
47	342									
33	211									
33	573									
26	616	430	3 920	3164	692	52	12	. .	. .	. .
53	255									
55	44	147	913	726	159	. .	14	14	. .	. .
79	49									
06	152									
59	316	156	1 363	689	632	42	. .	. .	. .	. .
20	191									
05	195	192	800	237	452	37	62	. .	. .	12
3	190	190	248	154	85	9	. .	. .	. .	. .
99	. .	. .	10 186	7500	2319	168	131	28	14	26

(c) Production of Eggs and Fry

Of all fish killed 146 males and 159 females were identified with certainty, which was assumed to represent a 1 : 1 sex ratio. All fish of III group and older were sexually mature and, even among I group fish, 80% and 91% respectively were identifiable as males and females.

The plot of egg numbers on fork length of females gave a linear regression (see also Nicholls, 1958c) described by the formula

$$N = 61{\cdot}24\ (L - 15{\cdot}65)$$

where

N = number of eggs
L = fork length of female

Egg production was computed from this relationship for mature females for each zone of the Horokiwi during 1940 and 1941. Table 8.6 shows a sample of the results for the 1938 year-class. Other year-classes for which these calculations were made ranged from

Table 8.6
(After Allen, 1951)

The number of mature females and the computed egg production for the 1938 year-class in each zone in the year 1940

Year-class	Zone	No. of Mature Females	Mean Length (cm)	Average Egg Content	Total Egg Production
1938	I	135	33·0	1060	143 100
	II M	85	30·0	880	74 800
	II R	18	30·0	880	15 850
	III	191	28·0	760	145 150
	IV	113	23·5	480	54 200
	V	7	27·5	730	5 100
	Total ..	549	..	..	438 200

1939 to 1936 inclusive. Allen was able to arrive at the following values for the river as a whole: in 1940, 989 mature females produced 888 550 eggs; in 1941, 1223 females produced 886 250 eggs. The final error involved in the estimated egg production for the whole population was apparently about ±50%, an estimate derived from consideration of three sources of error involved in the determination of the sex ratio, the size structure of the population and the proportion of the (female) population reaching maturity at a given age.

The number of redds detected in the river was markedly less than would have been predicted on the basis of total egg production and the average number of eggs per redd determined in other investigations. Allen attributed this discrepancy to the use of the same gravel area for redd construction by several females, to destruction of redds by flood and to simple inability to identify them. In 1941, Allen found that, of 11 marked redds, nine were destroyed by floods between late June and early September.

(d) Mortality

Figure 8.4 shows survivorship curves for three zones. For the Horokiwi trout Allen considered that a high early mortality rate led to a moderately low mortality rate succeeded by a rather higher rate for the remainder of life. The first year of life, which included the high post-hatching mortality about which there was little direct information, gave a survival of about 28% ($m = 0{\cdot}0039$ if time t is

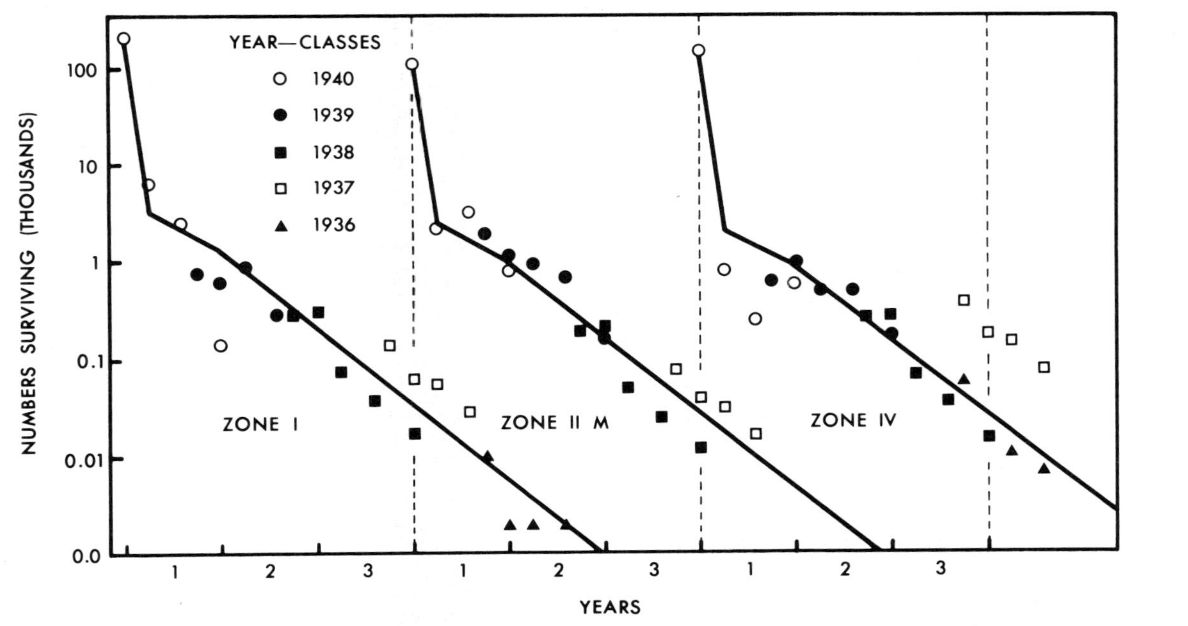

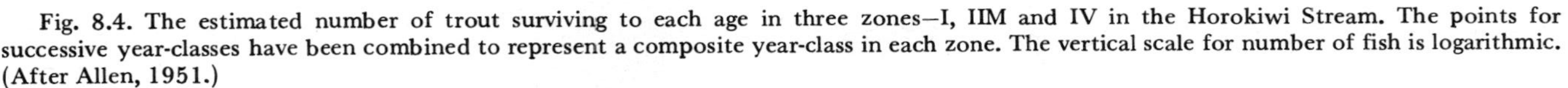
Fig. 8.4. The estimated number of trout surviving to each age in three zones—I, IIM and IV in the Horokiwi Stream. The points for successive year-classes have been combined to represent a composite year-class in each zone. The vertical scale for number of fish is logarithmic. (After Allen, 1951.)

given in days). The average slope of the survivorship curve was higher after the first year, giving $m = 0{\cdot}00486$ (17% survival of original population).

(e) Growth

Two methods were employed for growth determination: repeated measurement of the size of members of particular year-classes (identified from their scales) in the different sections of the stream and direct assessment of the growth of marked individuals. The methods checked fairly well against each other. One interesting demonstration was that growth differed considerably in different zones, which was yet another instance of the strongly conservative behaviour of the Horokiwi trout population.

Weight growth generally corresponded to length growth and Allen computed instantaneous growth rate as

$$\frac{\log_e W_2 - \log_e W_1}{365}$$

see Chapters 2 and 4. The mean instantaneous growth rate plotted against age of fish (Fig. 8.5) shows a very pronounced drop over the first year of life.

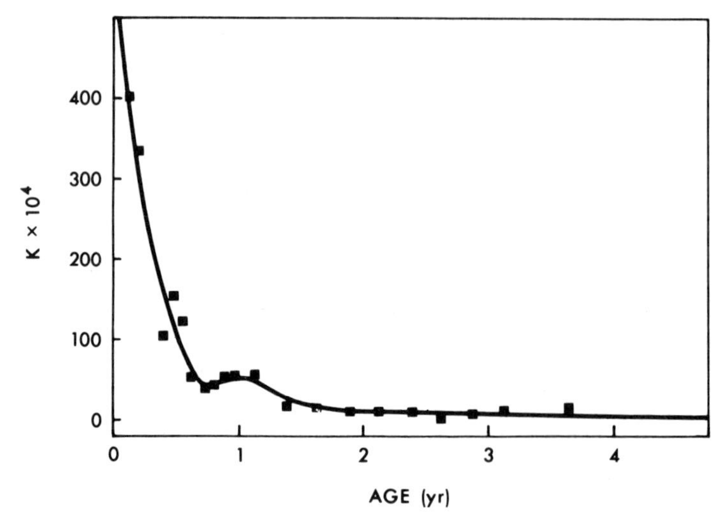

Fig. 8.5. The mean instantaneous growth rate ($\times 10^4$) plotted against age of trout in the Horokiwi Stream. (After Allen, 1951.)

(f) Food

Stomach contents of 380 fish of the 1939 year-class and of 81 older fish were examined. Forage ratios (percentage of a food item in stomach contents divided by its percentage in available fauna) were calculated for the most important types of food.

The contribution made by terrestrial organisms and fish to the food consumed was slight and in the main the food of trout in the Horokiwi was the benthic stages of four genera of Trichoptera and one or two genera of Ephemeroptera.

(g) Stock, Production and Crop

Stock, in Allen's terminology, equals biomass. From multiple samplings he was able to construct a picture of biomass change in the life of "composite" year-classes for both the upper and lower waters of the Horokiwi. That for the lower waters is given in Fig. 8.6. It is

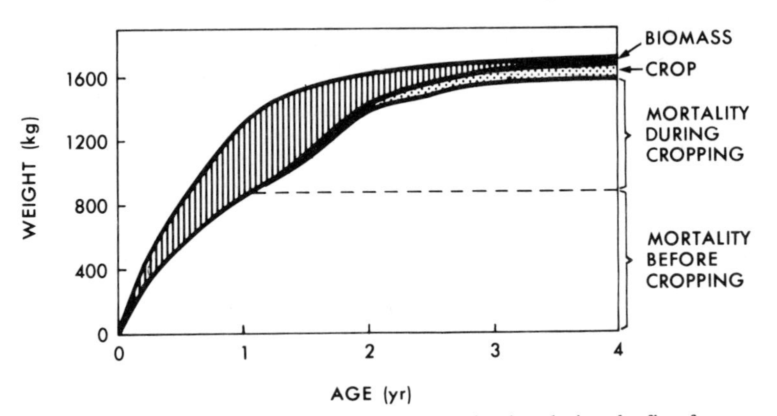

Fig. 8.6. The development of the accumulated production during the first four years of the life of a normal year-class in the lower waters of the Horokiwi Stream, showing the amount surviving as biomass, the amount that is removed by angling, and the amount dying from other causes. (After Allen, 1951.)

clear that the biomass of a year-class reaches a maximum in the second year of life. In other words, the product of fish numbers and mean weight is then maximal. Fish are much more numerous when younger which more than compensates for their being, on average, considerably larger when they are older, because then they are much less numerous. Figure 8.6 also shows cumulative production, which Allen obtained from the use of his graphical method (mentioned above). A typical use of the method appears in Fig. 8.7.

Biomass, expressed as a percentage of accumulated production demonstrates several interesting points, of which the most important is that changes in biomass accounted for only a trivial fraction of production in the older age groups. More generally, over 90% of the estimated total production of a year-class was complete by the end of the second year.

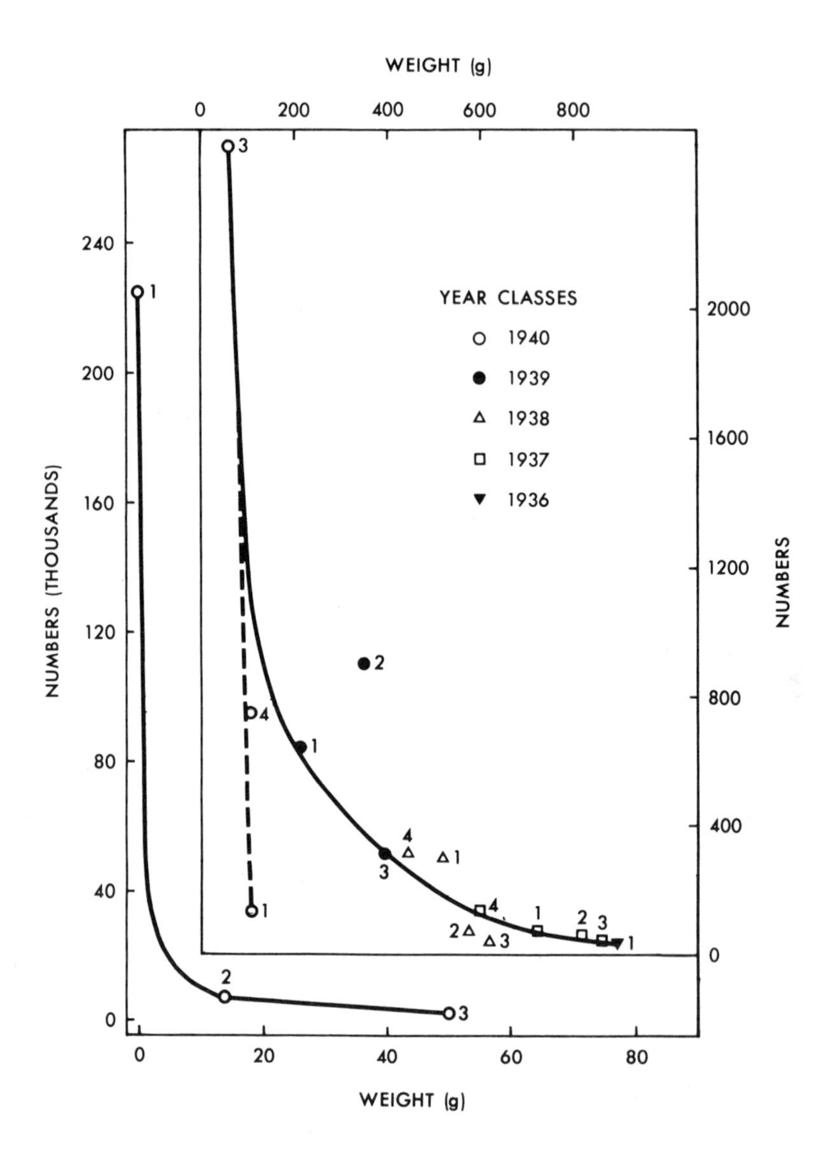

Fig. 8.7. The number of survivors plotted against mean individual weight at the same date for a succession of year-classes in Zone I of the Horokiwi Stream. The figures beside the points indicate the months for which the estimates were made; 1 = October, 2 = January, 3 = May, 4 = July. The broken line leads to a point for October, 1941, when the numbers appear to have been abnormally reduced by the floods. Owing to the wide variation in numbers the early and later parts of the curve have been plotted on separate scales. (After Allen, 1951.)

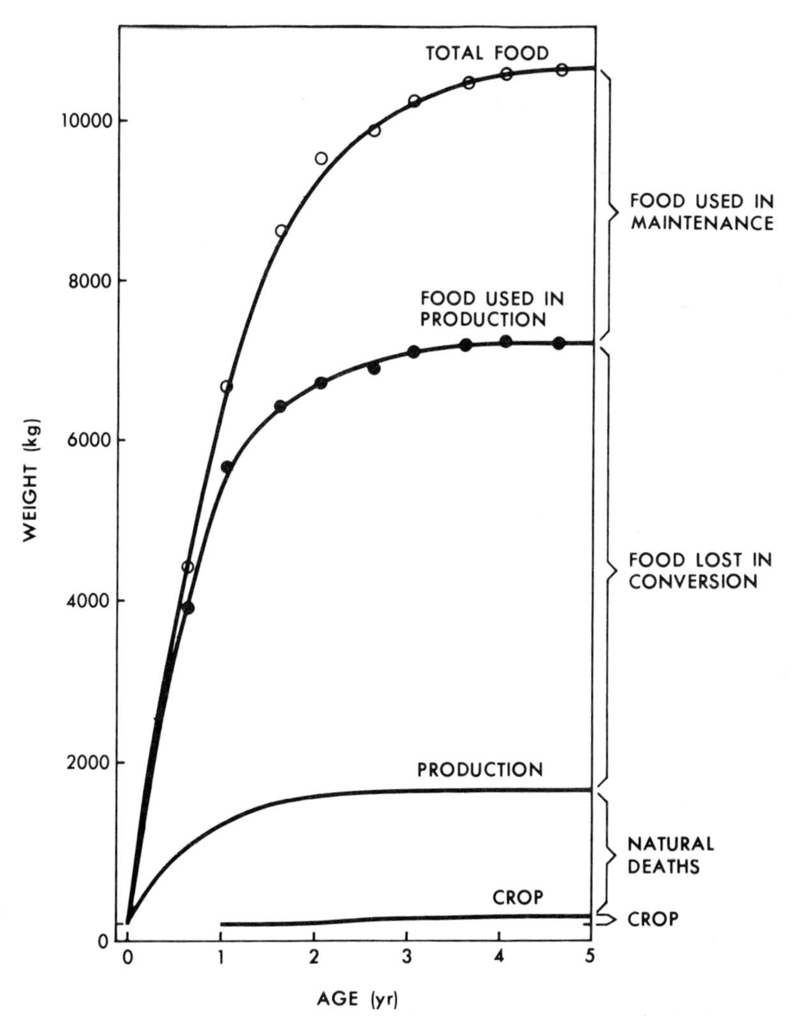

Fig. 8.8. Comparison during the life of a composite year-class in the lower waters, between total food consumption, the amount of food used in production, the weight of trout produced and the angling crop. (After Allen, 1951.)

Figure 8.8 depicts Allen's estimates of the total food consumed in a composite year-class in the lower waters of the Horokiwi as used in production etc.

C. Additional Production Studies

Gerking (1962) performed a notable investigation of production and food utilization by bluegills in Wyland Lake, Indiana. The essential natural history of his study was carried out in a manner reminiscent of that of Ricker and Foerster (1948) and Allen (1951), with the advantage of their recorded experience and of other more recently accumulated data. Gerking determined growth rates in Wyland Lake and, from calculations of exploitation rates derived from mark-and-recapture data, was able to determine that natural mortality was about seven times greater than fishing mortality. In accounting for food consumed, Gerking expressed it in terms of protein content, employing the data of earlier work (Gerking, 1952, 1954, 1955a, b) in computing the demand for protein for growth as observed in Wyland Lake bluegills. His calculations were certainly better than the earlier attempts by Allen (1951) to estimate food consumption rate by trout, because he had more complete knowledge of the relationship between rate of food (protein) utilization and weight of fish. Moreover, a knowledge of the relation between weight and length enabled Gerking to determine weight of protein consumed at different body lengths. Protein consumption of individual bluegills of different age groups in Wyland Lake, calculated on the basis of mean weight of fish, matched well with values derived from mean lengths (about 5% difference between means). The major sources of error Gerking (1962) mentioned derived from differences in temperature and activity levels in Gerking's (1952, 1954, 1955a, b) experiments as compared with Wyland Lake. Gerking considered temperature to be the more important of these differences, but the basis of this supposition seems tenuous. In his own paper (Gerking, 1962), he cited the observations of various workers which show that diel movements of fish within water bodies may be of considerable magnitude (e.g. yellow perch—Hasler and Bardach, 1949; Carlander and Cleary, 1949; Bardach, 1955; Hergenrader and Hasler, 1967).

Gerking attempted to determine rates of production by invertebrates in Wyland Lake from consideration of the feeding habits of bluegills as derived from their stomach contents, supplemented by direct sampling of the Lake's plankton, phytomacrofauna and benthos. By determining the seasonal changes in biomass of these animals and computing their protein values (derived from Geng, 1925) and coupling this with computations of seasonal food (protein) demands of the bluegills, Gerking made estimates of the production rates of invertebrates available as food required to sustain the observed levels of bluegill production.

Gerking (1962) also re-examined similar calculations made earlier by Allen (1951) for the fauna of the Horokiwi. He concluded that Allen's estimates, that the Horokiwi produced about 100 times its biomass to keep up with depradations by trout, were two or three times too high. He showed that Allen had not been in possession of suitable data to allow for the probable effect of size of fish and of temperature on maintenance needs, and that these lacks probably constituted two major sources of error in Allen's calculations.

Gerking's own excellent study enabled him to identify two distinct periods in the growth or production pattern of organisms comprising the food of bluegills (Gerking, 1962). Bluegills in Wyland Lake grow—and therefore produce—for only five months in the year, whereas benthic production extends over 12 months. Fish consume so much food while growing in summer that the benthic production, although high at that time of the year, is inadequate to compensate completely for removal by fish; this is what Gerking termed the "keep up" phase of benthic production. Benthic biomass would consequently decline to zero eventually, except for the seven-month period during which fish do not grow (or produce) and therefore make minimal food demands. This is the "catch up" period for the biomass of the benthos. In sum, Gerking found that Wyland Lake's production of bluegills was about 91 kg/hec and that annual production of those animals available as food was about 10 times greater.

The above study has two apparent defects. Firstly, like other such field studies, it lacks information on the food demands for activity by wild fish. Secondly, in spite of Gerking claims to the contrary it is far from certain that protein consumption or assimilation is a particularly good production index. Gerking's rationale is that body fat content is unpredictably variable, that carbohydrate is stored in the body only to a negligible extent and that protein is the least variable and generally most abundant constituent of protoplasm (water aside). However, rapid deposition of fat or increase in its concentration in tissues seems to be an important aspect of growth phenomena in fish (Nikolskii, 1969; Le Cren, 1962), especially in circumstances where food is particularly abundant or where fish are preparing physiologically for breeding or to undergo a major migratory movement. It is therefore difficult to accept that a consistent adherence to protein as an index of growth or production can be without potential computational shortcomings.

Chapman (1967) has pointed out that the ratio of annual production to average biomass in fish populations frequently falls within fairly narrow limits. And such data as those of Ricker and Foerster (1948), Allen (1951), Gerking (1962), Patriarche (1968), Kipling and Frost (1970) and Egglishaw (1970) do lend support to

Chapman's idea that, at least in some cases, natural regulatory factors tend to control biomass (see also Chapter 6). Clearly, if populations are in steady state with food production relatively constant in pattern, we can expect that production and biomass of various age/size groups will bear roughly constant relationships to each other at various times of the year. And, for the year as a whole, the ratio of average production to average biomass could also be expected to bear a constant value from year to year. An appreciation of this could be reasonably important in making crude predictions of production from biomass in certain instances. Carlander (1955) published data on fish biomasses in North America which showed that these were highest in river backwaters and oxbows, reservoirs and ponds. Warm water lakes and trout lakes (cooler water) had much lower biomasses. Whereas area was not significantly related to biomass, depth and carbonate content of water were inversely and directly related, respectively, to biomass. Carlander found that species of fish with short food chains had the highest biomasses, those with food chains of intermediate length had biomasses of intermediate values and the "secondary predators" with long food chains had low values. Carlander cited this relationship between biomass and feeding habits as a demonstration of the "pyramid of numbers" (Elton, 1927). It is certainly interesting to realize that the food habits of fish—somewhat inexorably tied, as they are, to functional morphology (Chapter 5)—do probably control their trophic level and therefore limit the range of biomasses (and probably production values) they are *able* to manifest in natural systems. Figure 8.9 is adapted from Carlander (1955).

Apart from such possible extensions of the production : biomass ratio as this, it is difficult to see that any higher generalizations about it can be made out at this time. After all, during the course of a year the ratio will usually be changing constantly (Ricker and Foerster, 1948; Allen, 1951; Egglishaw, 1970), as Figs 8.3 and 8.6 clearly show. (Figure 8.6 also shows that biomass of a year-class tends to reach a peak early and decline later due to mortality. Egglishaw (1970) indicates the same trend.) The bulk of production that a year-class is capable of usually occurs in the early part of its life. Later work by Gerking (1962), Le Cren (1962), Kipling and Frost (1970) and Egglishaw (1970) tends to the same conclusion.

In Table 8.7 calculations of production in tench living in Tasmanian farm dams are given, based on data of Weatherley (1958). The fish were introduced into these dams—formerly without any fish—as young, one- or two-year-old, individuals. By repeatedly

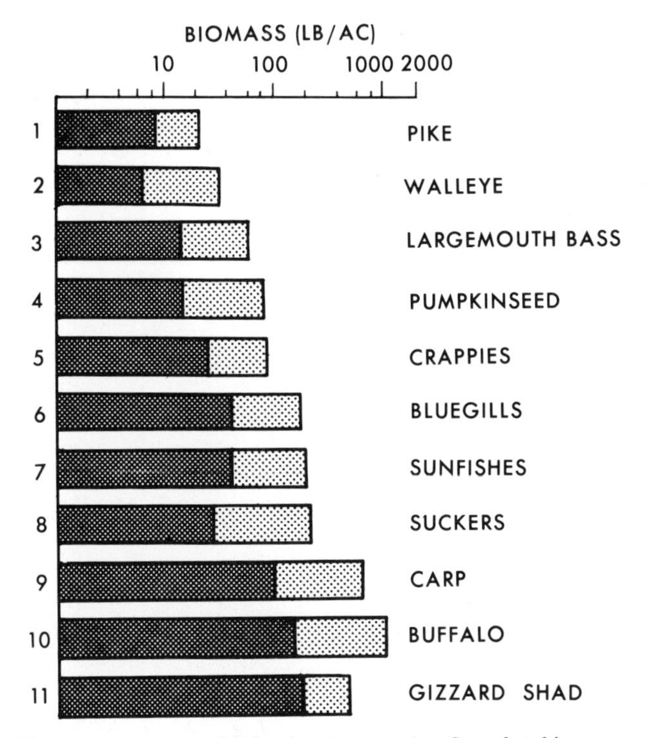

Fig. 8.9. Biomasses per acre of fish of various species. Cross hatching represents mean values for lakes and reservoirs; stipple represents maximum values. (Simplified after Carlander, 1955.)

seining these dams on subsequent occasions (the dams are small, smooth-bottomed circular or oval depressions) it was possible to catch nearly every fish surviving. On the basis of observed growth of the survivors and knowing mortality, it was a simple matter to compute (i) the production due to biomass increase and (ii) the production of those fish that had died, between samplings, according to the equation given earlier where

$$P = B_1 - B_0 + B_d \text{ (Chapman, 1967, 1968).}$$

From this the production in pounds per acre is shown for tench in each of eight small farm dams. The point of interest in Table 8.7 is that the relative contribution, by fish that die, to total production is higher where mortality is greater. This sort of principle is the one that prompts so many fishery biologists to recommend predator control as a way of diminishing mortality and therefore reducing that aspect of production otherwise unavailable to man.

Table 8.7

Production, biomass and survivorship and cumulative mortality of tench in eight Tasmanian farm dams (Based on data from Weatherley, 1958)

Dam No.	Production Period	Biomass Gain (lb/ac) B	Production of Fish that Die P	$B + P$	$\frac{P}{B+P} \times 100$	Surviving (%)	Cumulative Mortality (%)
1	29.9.54–17.7.55	103·6	8·5	112·1	7·6	85·9	14·1
	–14.3.56	19·5	0·6	20·7	2·9	80·9	19·1
	–7.11.56	41·9	10·7	52·6	20·3	53·5	46·5
2	29.9.54–17.7.55	38·7	8·0	46·7	17·1	70·5	29·5
	–14.3.56	20·4	0·0	20·4	0·0	70·5	29·5
	–7.11.56	25·7	7·3	33·0	28·4	44·9	55·1
3	3.12.55–20.4.56	36·2	13·1	49·3	26·6	58·0	42·0
	–16.4.57	2·4	0·07	2·47	2·8	55·0	45·0
4	3.12.55–12.3.56	51·7	2·0	53·7	3·7	92·9	7·1
	–14.12.56	57·7	12·3	70·0	17·6	65·2	34·8
	–16.4.57	40·6	0·0	40·6	0·0	65·2	34·8
5	21.9.56–24.4.57	7·7	4·8	12·5	38·4	44·4	55·6
6	21.9.56–24.4.57	27·4	2·8	30·2	9·3	83·3	16·7
7	21.9.56–25.2.57	38·0	6·3	44·3	14·2	75·0	25·0
8	21.9.56–24.4.57	2·3	5·2	7·5	69·3	18·2	81·8

Le Cren (1962) has drawn attention to the often neglected component of fish production that goes into manufacture and release of gonad products—ova and sperm. He showed that this could account for nearly half of total production by adult fish, in the case of perch, and although this aspect of production can be much less it can frequently assume extremely important proportions.

Le Cren (1962) also suggested that, to some considerable extent, production of fish was determined by the population density. That is to say, given the flexibility of growth rate observed, a certain minimum number or biomass of fish must be present initially before a population or age class can display its maximum production rate in a given environment. In later developments of the experiments with young trout on which these claims are based Backiel and Le Cren (1967) showed that the general form of this relationship was determined within 40 days of the commencement of feeding by fry and that a quite critical population density (8 fish/m^2) could be identified below which the potential for production could not be fully realized. Other experiments, with carp *Cyprinus carpio,* were cited which show that the maximum production is not always asymptotic with respect to density, but is sometimes found at "moderate" population densities.

II. GENERAL COMMENTARY

This chapter began with a generalized introduction to production measurement in fish populations which led to relatively detailed recountings of case studies of production. Any lessons drawn from these depend on the principles which they exemplify and in the practical problems that their authors overcame.

In the first place, it is obvious that these studies were conducted in places where natural environmental advantages simplified their execution, but that is not to say they were technically easy. Allen's investigation, in particular, represented a huge amount of arduous field work undertaken mainly by himself and one assistant. But the studies were based on populations in situations in which the whole life cycle, or a large and important part of it, was continuously present within one water body. The same may be said of the recent study of pike populations in Windermere (Kipling and Frost, 1970).

The O group of Cultus Lake salmon was much more numerous than older groups and, so far as production is concerned, easily the most important group. That most of the salmon left the lake when one year old, was actually advantageous to Ricker and Foerster because it provided them with a ready means for censusing stocks. The relative sameness of growth and mortality patterns within the lake made it feasible to apply quantitative findings for just a few years to the various years of the study, with appropriate adjustments. Other species preying on the sockeye presented Ricker and Foerster with a most useful means for assessing growth of the latter over short periods by examining the stomach contents of the former.

In the Horokiwi Allen not only enjoyed the advantages of studying a small, self-contained ecosystem of which trout were a part, but was fortunate in being able to treat various sections of the river as discrete units, although this did make it relatively difficult to obtain population estimates for the river as a whole. Allen was, however, able to compare growth rates in different parts of the river because the trout were so conservative in their movements that local differences in growth rate, probably the results of different feeding conditions, were apparent.

These studies are important in particular because the authors had understood the nature of production processes in theoretical terms. A great deal of work was still necessary to obtain the data on which to base actual estimates of production in natural populations; in the case of salmon the estimates were applicable only to the early landlocked phases in the life cycle and in the trout a serious error in the computations of production was the omission of the values for egg production—even though these were actually determined in part.

While it is clear that egg production, recruitment success and the survivorship patterns that apply throughout life all profoundly influence production it is also apparent that production is very closely tied to growth. In a major sense production *is* growth, the growth achieved by those that die as well as the growth of those still alive at the time of its assessment. Without again entering into the analysis of growth of the individual fish attempted in Chapter 4, one must obviously obtain quite accurate measurements of growth in the study of production in populations, often going to very considerable pains to secure the data. From the work examined in this chapter, especially Allen's, and from what has been discussed in earlier chapters, it seems apparent that growth, as one of the most labile aspects of fish physiology, may be expected to respond sensitively and rapidly to changes in environmental factors such as temperature and food. Growth, then, and the factors which govern its rate must lie at the heart of any investigation of production in fish populations.

A number of investigators of fish production have made much of estimates of "maintenance" food requirements of fish, so that they may add this value to the observed food requirement for growth and thereby determine how "efficiently" fish are utilizing the food available to them. The assumptions they have been required to make about what constitutes a reasonable value to assign to "maintenance" needs seems to be one of the major systematic shortcomings of production studies so far. As was stressed in Chapters 2 and (especially) 4, the need for hard data on the food requirements of normal fish activity in the field in different age/size groups and at different temperature levels has now become quite crucial for genuine further progress in the analysis of production processes.

9 | The Trophic Environment and Fish Growth

Most studies of fish population dynamics and growth have been restricted to determining responses of a population to changes in certain features of the environment. A limited number of investigators have, however, taken the view that every fish population is part of a *system* (ecosystem), affected by chemical and physical factors while functioning within the system through biotic—especially trophic—interrelationships. A few examples of this sort will be considered in this chapter.

I. THE SCOTTISH SEA LOCHS EXPERIMENTS

Cessation of the North Sea fisheries during 1939-45 stimulated certain investigators to experiment on sea lochs, to determine whether their primary productivity might be increased by enriching them with chemical fertilizers. It was hoped that if chemical enrichment led to augmented food chains this would result in improved growth and production of fish—a standard result in those parts of the world where fish are cultured in ponds for food production or recreation. Such fish ponds are usually endorheic and hydrologically and biologically rather simple systems. The addition of inorganic or organic fertilizing substances of either natural or artificial origin usually results in a successful elevation of production. It is, of course, necessary to plan carefully and carry out the fertilizer or enrichment procedures in a competent manner to ensure success but the world-wide presence of fish culture industries indicates that this is perfectly feasible (Swingle and Smith, 1942; Mortimer and Hickling, 1954; Hickling, 1961, 1962; Schaeperclaus, 1933). It is, however, quite another thing to place fertilizers in an extensive natural water body of a complex and unknown hydrological and biotic character and subject to continual loss or exchange of its original water mass. The experiments on Scottish sea lochs were therefore pioneer studies from the fisheries standpoint and also provided an illuminating example of the movement and biological fate of nutrients within certain types of marine or semi-marine situations.

An initial experiment was performed on Loch Craiglin and a further one on Kyle Scotnish, two lochs on the west Scottish coast, both connected to the sea through Sailean More. Loch Craiglin is much influenced by freshwater runoff, being connected to the sea only over a shallow sill, whereas Kyle Scotnish is a true arm of the sea. Attempts were made, which were complicated by the daily effects of tides in Kyle Scotnish (Nutman, 1950) and by a periodic rainfall in the case of Loch Craiglin (Orr, 1947), to describe and measure physical and chemical conditions in the lochs.

A. The Loch Craiglin Experiment

Loch Craiglin, only 18 acres in area, was chemically fertilized as follows:

Year	Superphosphate (lb)	Sodium Nitrate (lb)
1942	319	515
1943	301	549

The fertilizers were added with reasonable regularity as roughly constant fractions of the total quantities. A smaller amount was also added early in 1944. In all, the loch was fertilized 11 times over a period of 20 months.

The μ flagellates of the phytoplankton responded most rapidly to chemical enrichment. Numbers increased at all depths two to three days after each fertilizer addition. But if nutrient utilization was rapid it was also short-lived as the "pulses" of increase lasted only briefly. Among diatoms and dinoflagellates increases were delayed until autumn when densities of these organisms in Loch Craiglin considerably exceeded those in outside Sailean More (Marshall, S. M., 1947). In explaining why densities were not even higher Marshall pointed out that increases in phytoplankton abundance could be somewhat offset by accompanying increases in zooplankton grazing activity.

Of the zooplankton, rotifers and larval molluscs reached greater densities in the loch than in Sailean More but copepod populations were more variable. The highly alkaline conditions (pH 9) that followed enrichment may have been deleterious to some zooplankton (Marshall, S. M., 1947).

In the benthos, very little change showed during the first year of the experiment but there was a massive increase later, mainly of *Cardium* and *Hydrobia* and of chironomid larvae (Raymont, 1947). Estimates were made of the benthic animals fish could consume, compared with those that were themselves competing with fish for food on the remainder of the benthos. Even after due allowance for this, it seemed evident that benthic food for fish was two or three times more abundant than before chemical enrichment (Figs 9.1 and 9.2).

It was difficult to evaluate the effects of the augmented benthic food supply on fish growth. Before enrichment Loch Craiglin contained seven species of fish. Six hundred small plaice were added

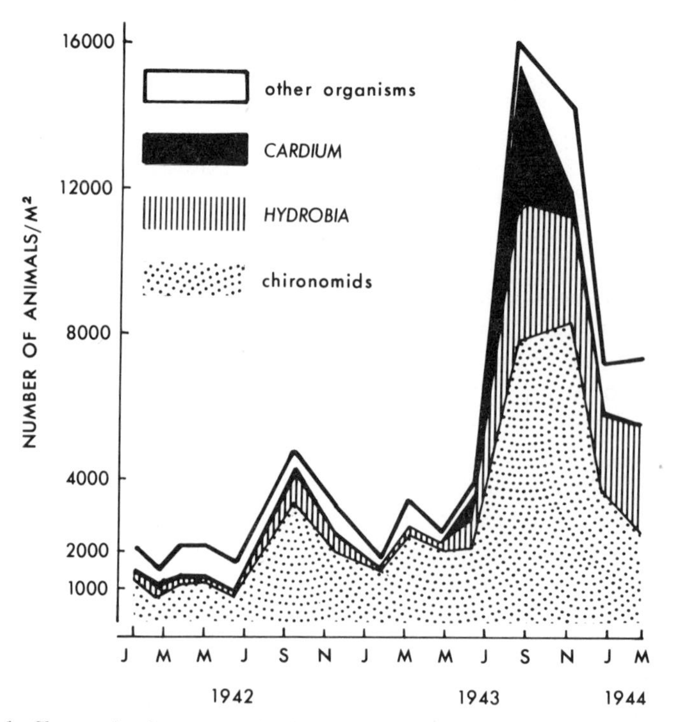

Fig. 9.1. Changes in the average density of benthic fauna in Loch Craiglin. Each value represents an average estimated from the populations at five regular sampling stations. Shading indicates the more important groups comprising the fauna. (After Raymont, 1947.)

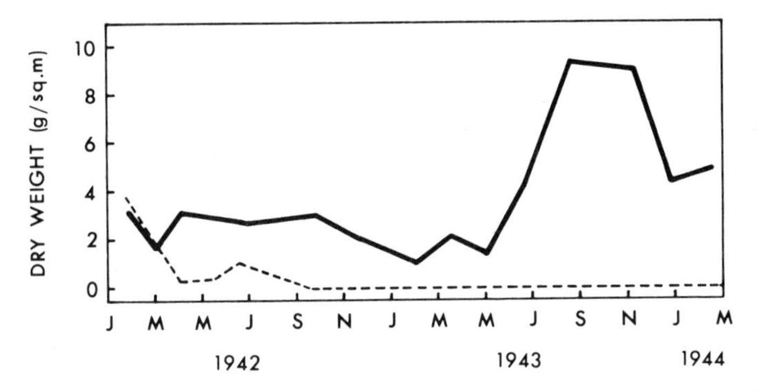

Fig. 9.2. Changes in the dry organic weight of the benthic fauna. The total weight per square metre of species useful as food for fish (full line) has been separated from that of "competing" species—*Ciona* and *Anemonia* (broken line). Each weight was obtained by multiplying the average population per square metre for each species, estimated from samplings at five regular stations, by the average dry weight of the flesh of each species. (Modified slightly after Raymont, 1947.)

to these in 1942, of which 425 were marked, plus 2600 small unmarked flounders. The loch had previously lacked flounders (though not plaice) and their subsequent growth was compared to that of flounders of the same age in Loch Killisport from which the Loch Craiglin stock had come. The growth of the flounders in Loch Craiglin eventually exceeded that in Loch Killisport by four times in length and 16 times in weight over the period July 1942 to April 1943. Between April and May 1943, about 1100 flounders were added to Loch Craiglin and in July of the same year more than 21 000 O group flounders. These three groups of flounders released in Loch Craiglin were termed Stocks I, II and III respectively (Gross, 1947).

From April to November 1943 Stock I flounders grew 2½ times more in length and 18 times more in weight than their Loch Killisport counterparts, attaining in less than two years a size reached in Loch Killisport only after six years. Stock II flounders grew four to six times faster (in terms of size-for-age) than Killisport flounders and Stock III also show a markedly greater growth rate.

In one year in Loch Craiglin the unmarked plaice reached a size normally requiring two years. However, the growth of marked plaice was no better than average and marking evidently caused some retardation of growth (Gross, 1947).

The results on flounders, though very informative, failed to furnish conclusive evidence of increased growth of fish due to an augmented food supply. It could, for instance, be argued that conditions in Loch Craiglin were especially suitable for fish growth for reasons unconnected with chemical fertilization.

B. The Kyle Scotnish Experiment

Apart from the different environmental setting (Kyle Scotnish is about one mile long) and the important effects of tidal movement in which some 8% of the volume of water in the loch is exchanged daily with the surrounding sea (Nutman, 1950), together with the more typically marine environment, the results of the Kyle Scotnish experiment were similar in general trend to those of Loch Craiglin. Additions of fertilizer were again followed by short-lived but considerable increases in phytoplankton. Zooplankton abundance followed a seasonal course similar to that in Sailean More, but at an appreciably higher level (Gauld, 1950). The benthic changes again suggested a major increase during the first 18 months after the beginning of chemical enrichment, building up to a density maintained thereafter—at least for several years (Raymont, 1950).

Growth of plaice and flounders introduced to the loch greatly exceeded that on nearby fishing grounds (Gross, 1950). However, the same criticisms apply to this experiment as to the Loch Craiglin one, since it could again be argued that Kyle Scotnish might be naturally very favourable for fish growth.

C. Commentary on the Scottish Sea Loch Experiments

These experiments were very interesting pioneer studies that excited the attention of ecologists and hydrobiologists, but they suffer from some undeniable deficiencies. It seems fairly certain that phytoplankton underwent great temporary increases following many of the chemical enrichments, an observation which applies generally to nearly all chemical enrichment work, whether of marine or inland waters (Swingle, 1947; Schaeperclaus, 1933; Mortimer and Hickling, 1954; Hickling 1961, 1962; and see Table 9.1).

The evidence for increases in zooplankton was less decisive, but there appeared to be definite increases in the benthic fauna.

The disappointing feature of the work and a lesson for fish ecologists is that there was no formal preliminary investigation of fish growth. This could have taken the form of a study of growth in the indigenous populations, which contained plaice in Loch Craiglin and small numbers of both plaice and flounders in Kyle Scotnish, by means of examination of scales and otoliths, with growth before and

Table 9.1

Examples of the biological effects of artificial enrichment of lakes

Name or Location of Lakes	Effects					Remarks	Authorities
	Phytoplankton	Zooplankton	Bottom Fauna	Plants	Fish		
Wisconsin	?	+	?	?	+	No effect until soyabean meal added to fertilizers	Juday, Schloemer and Livingstone, 1938
Crecy Lake, New Brunswick	+	+	+	+	+	Significant increases in fish yield only after intensive predator control	Smith, M. W., 1945, 1948, 1955
Ontario Lakes	+	?	?	?	?	Inconclusive	Langford, 1948
Michigan Lakes	?	?	?	?	+	Heavy "winterkill" of fish and invertebrates under ice in enriched lakes	Ball, 1948
Three Dubs Tarn, England	?	?	+	?	+ (?)	Bottom fauna increased ten-fold. Fish probably affected, but effects unsustained	Macan, 1949; Frost and Smyly, 1952
Bare Lake, Alaska	+	+ (?)	?	?	+	Rate of rotifer production increased; no apparent increase in crustaceans	Nelson and Edmondson, 1955
Lake Dobson, Tasmania	+	+	?	+	+	Great increase in plants with associated fauna	Weatherley and Nicholls, 1955
Loch Kinardochy, Scotland	+	?	+	?	?		Brook and Holden, 1957

+ = Increased abundance or growth rate. ? = Doubt as to effects. (?) = Effects not demonstrated decisively.

after enrichment compared. Alternatively, or in addition, stocked fish could have been marked and permitted to grow for some time in the lochs before enrichment was begun. Subsequent stimuli to growth might then have been revealed in a major shift in the course of average growth rate. Wartime difficulties and the need to secure rapid results were probably the cause of most of the departures from the ideal in this part of the experiments.

The most compelling evidence in support of an actual stimulus to growth is that both the chemically enriched lochs contained very rapidly growing fish as compared to those in the adjacent marine areas including other sea lochs.

II. OTHER CHEMICAL ENRICHMENT STUDIES

Table 9.1 indicates the essential biological results of various other similar projects many of which derived their inspiration from the Scottish sea loch experiments. Only one of these is described at length.

III. CHEMICAL ENRICHMENT OF A TASMANIAN TROUT LAKE

In their experiment on Lake Dobson, Weatherley and Nicholls (1955) added a commercial mixed fertilizer (consisting of 60 parts of superphosphate, 40 of ammonium sulphate, 15 of ground limestone and 5 of potassium chloride), initially in paper bags, later by broadcasting it from a rowing boat.

Phytoplankton and zooplankton were both increased at least temporarily, but there was also a more permanent effect of fertilizer—a major increase in growths of the plant *Myriophyllum elatinoides.* This formerly sparse and scattered plant grew into an unmistakable pattern of large clumps that accurately indicated where paper bags of fertilizer had penetrated the soft lake mud; many other *Myriophyllum* plants grew thickly in lines that traced the course of the boat from which fertilizer had been broadcast. Evidence was adduced that phosphorus was the main stimulus to growth of *Myriophyllum.* Mud collected from around the base of one of the major clumps a year after a bag of fertilizer had penetrated the mud showed a concentration of phosphorus of about 1000 μg/g silt at a depth of 1 ft below the surface. Concentrations of phosphorus fell off rapidly both with depth and distance from this point. At depths of 2 and 3 ft below the surface at the base of the clump, and at depths of 1, 2 and 3 ft at distances of 3 and 6 ft from the base phosphorus values ranged from approximately 100 to 200 μg/g silt. An analysis of plants growing in unenriched and enriched clumps, in which the concentrations of 10 elements (plus crude protein, ash and water content) were measured, showed phosphorus to be 3½ times higher in the enriched clumps than in the unenriched. The *Myriophyllum* was inhabited by a large epifauna dominated by gastropods and stoneflies distinct from the ordinary benthos, which may or may not also have been increased. However, the vast increase in plants must have represented a considerable increase in available trout food in terms of its epifauna.

Of 467 tagged yearling brown trout released in Lake Dobson in 1952, after enrichment had produced its major effects, 30 were subsequently recaptured and their growth while at liberty assessed by direct measurement and by back-calculation from scales. It was, however, back-calculation of growth of trout already present in the lake before enrichment that proved more interesting. Examination of scales of 16 such fish showed a fairly clear increase in their growth following enrichment (Fig. 9.3). The effect was the more notable

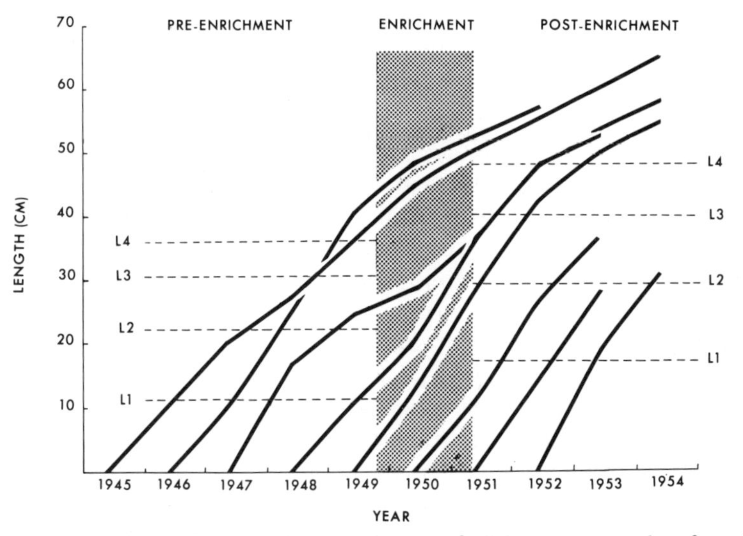

Fig. 9.3. Calculated mean lengths at each year of all brown trout taken from Lake Dobson. Horizontal lines indicate the mean lengths of all fish of the same age before and after chemical enrichment. The shaded column covers the period between the beginning and the end of the addition of fertilizers. (Simplified and modified slightly after Weatherley and Nicholls, 1955.)

because gill-netting in 1954 indicated that the survivors of the 467 fish added in 1952 had come to comprise about two-thirds of the total population by then. In other words, the increase in growth of resident trout during 1951-53 had occurred despite a very considerable increase in the size of the population. Four fish of the 1952 year-class, which were spawned naturally in the streams at the edge of Lake Dobson, showed as rapid growth in their first two years as any fish captured (Fig. 9.3). In this experiment there appears little doubt that available food was greatly increased—although through plant epifauna rather than benthos—and that trout growth increased as a result.

IV. FISH PONDS OF JAVA

Before Java became part of Indonesia Schuster (1952) gave a detailed account of its brackish water fish ponds, including many interesting observations on the life of those people who obtained their livelihood from a pond economy. At the time of Schuster's report the brackish water ponds—"tambaks"—were used mainly to culture the milkfish *Chanos chanos.* Since then, the cultural practices have in many cases been somewhat changed, but this does not invalidate or render less valuable Schuster's original assessment.

A. The Tambak Environment

Tambaks are constructed on alluvial mud flats on or near the sea coast in association with estuaries. Their basic bottom deposits are from rivers rather than the sea and under the influence of the constant submersion that follows construction of embankments the tambak mud assumes a form peculiar to the environment. As Schuster (1952) wrote, "The upper layer of tambak soil acquires the condition of a more or less jelly-like mud. This layer of soft mud should be recognised as the really productive zone of the pond." And this apparently remains true irrespective of the nature of the remainder of the tambak environment which, despite the relative thermal constancy imposed by a tropical climate, is very dynamic. The changeable nature of conditions derives from the small daily and large seasonal tide movements, which largely govern the amount of water available to fill tambaks. The variability of this supply, coupled with major seasonal differences in rainfall, can produce salinity variations from 18-46 per mille near the sea shore and from 0-2 per mille in tambaks situated three miles inland.

The mud, however, is the important feature in terms of fish growth, because it is on this that the algae grow which form the major component of the food of *Chanos*. The two principal algal groups as far as fish are concerned are the Myxophyceae and the Chlorophyceae. The conditions for their growth are defined by edaphic differences in tambak environments. Schuster (1952) described the edaphic characteristics for the two groups.

Mxyophyceae prefer	Chlorophyceae prefer
A soft, hydrophilic, biologically active mud.	A more or less solid soil.
Much organic matter in the soil.	An adequate nitrate and phophate content in the water.
A water level from 1-50 cm.	A water level from 30-100 cm
An oscillating water level or gently flowing water.	Stagnant water.

Myxophyceae typically occurred to 10 cm depth in mud, sometimes considerably deeper. Chlorophyceae were more confined to the surface of the mud.

Though milkfish can eat and grow on either type of alga there are certain complexities in their relationship to their algal food. According to Schuster Myxophyceae multiply continuously, ensuring a constant food supply, whereas Chlorophyceae offer fish little readily digestible food during the development of the algae but abundance when they become overripe and begin to decompose.

In further substantiation of this claim there is fairly convincing evidence that Chlorophyceae are less satisfactory food than Myxophyceae when milkfish are small. Aquarium observations showed that small *Chanos* find fresh filamentous Chlorophyceae indigestible and that, to become acceptable, these algal masses must first be softened by partial decomposition. Yet Schuster noted a belief among tambak operators that *Chanos* could reach a large size only when Chlorophyceae were plentiful. He resolved this apparent paradox by explaining that once *Chanos* reached a certain size they could consume, and grow on, coarser food—such as filamentous Chlorophyceae. In practice, it has been found that 200 young *Chanos* per acre, growing in normal ponds containing mainly Myxophyceae, reach an average size of 12 oz in one year. Further growth follows usually only if the fish are transferred to ponds bearing dense growths of Chlorophyceae. The fish, which will then be big enough to use it as food, can proceed to capitalize on the greater basic productivity of the Chlorophyceae.

Because the types of soil and other conditions necessary for good growth of Myxophyceae or Chlorophyceae are definable, the outline of a control strategy for growth of whichever sort of alga is desired can be traced. The heart of a rational system of management lies in recognizing the role of the micro-vegetation of the pond bottom and the factors that maximize its growth. The growth of each of the two main types depends on the influence of the already described edaphic factors but not on others which, on *a priori* grounds, might easily have been inferred as of equal importance. Salinity changes are, for instance, not very significant nor is pH. On the other hand light, oxygen and properties of the mud are very important.

V. DISCUSSION

It can be seen at once that these case studies are very much on the lines of the "ideal" problems in ecology suitable for investigation by systems analysis, as outlined in Chapter 1. They were all carried out before ecologists consciously began to use the concepts and technical language of systems analysis and before large scale automated data collection and processing had become common. When work of a similar nature is performed today there are many more ways available for the rapid collection of bulk data, especially when groups of investigators are on hand for the work. In a sense, this does more than merely increase the amount of detailed information that can be collected: it transforms the entire scope of such problems.

For instance, many of the trophic relationships thought to exist in the studies described depend essentially on appearances—seldom on more than a certain strength of correlation. To illustrate this, let us consider the framework of ideas that grew up around the work in Lake Dobson. First, we can list the changes that followed chemical enrichment:

Treatment	Changes following fertilization Chemical	Biological
Inorganic fertilizers added	(i) Transient increases of P in water	(i) Increases in phytoplankton density
(i) in paper bags penetrating mud	(ii) Increase in O_2 concentration in water	(ii) Increase in zooplankton density
(ii) broadcast	(iii) High concentration of P around base of new weed clumps	(iii) Great increase in weed (*Myriophyllum*) clumps
	(iv) High concentration of P in substance of new weed clumps	(iv) Great accompanying increase in weed epifauna
		(v) Superior growth of trout

There was no doubt that fish grew considerably faster, nor did there seem any doubt that this could be related in an essentially causal way to chemical enrichment. This chain, or web, of relationships was thought of as having the approximate form given in Fig. 9.4. This figure reveals some possible relationships between growth and fertilizer through changes in the food web but some of these relationships are more tenuous than others. However, the scheme does give information in a form which could readily be converted to the "black box" diagram of the systems analyst. Any part of the system deemed worthy of more intense study could be detached from the whole system and examined by means of appropriate technique so that its internal relationships could be clarified. Moreover, if such a sectional study were undertaken its findings could still be located logically within the food web proposed for the entire system. Let us, for instance, consider the part of Fig. 9.4 surrounded by a dotted line. What do we know of a factual nature concerning this?

(i) Growth of trout in the lake increased significantly following chemical enrichment.

(ii) Plant epifauna (essentially stoneflies and gastropods), of a sort eaten by trout and readily available to them, was greatly increased because of the vast spread of plants.

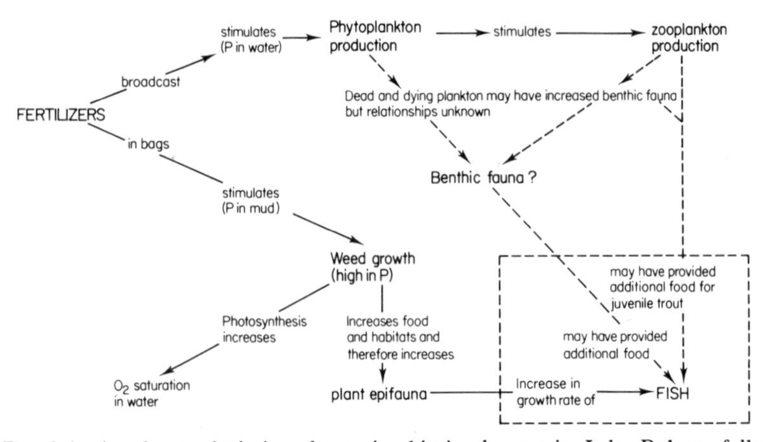

Fig. 9.4. A scheme depicting the major biotic changes in Lake Dobson following chemical enrichment. See text for further explanation. (Original, based on data of Weatherley and Nicholls, 1955.)

In addition we may infer that benthic fauna increased and that zooplankton was eaten by small trout in greater quantities than before enrichment—though there is no direct evidence for either of these inferences.

One thing seems clear: the increase in fish growth was due mainly to a greater food supply, because the release of 467 yearling trout in 1952 would certainly have increased competition for food. If food was not considerably increased during 1950-52, trout growth after 1952 could have been expected to be less, not greater, than previously.

The focussing of attention on this section of the problem would allow for study of the details of plant epifaunal production and effectiveness of its utilization by trout, together with estimates of conversion efficiencies. This could be made the subject of specific physiological experiment if desired. The presence of many age/size groups of trout, together with their possibly different states of mortality and food utilization, would invite further analysis of this part of the problem to whatever level seemed expedient or appropriate. The diagram of the relationships in Fig. 9.4 is, of course, a hypothesis or rather an interlocking series of hypotheses. The leads into the particular hypotheses about effects on trout growth within the boxed area can be traced backwards along their courses into the other sections of the scheme. These other sections may themselves be boxed, arbitrarily, and considered as discrete (though interconnected) parts of the whole system. In that way their precise relationships to the growth of trout can be examined while they are, of course, simultaneously being tested themselves for inner consistency.

This approach to questions of growth—and for that matter of other ecological problems—leads to no complete "final solutions". Rather does it subsume limited final solutions, such as questions of how efficiently an animal is able to convert food to flesh within a particular environmental setting. The method enables the investigator to build up a detailed stategic map of an ecosystem. Like other maps, this one will not everywhere be equally replete with detailed information. It will highlight, emphasize and magnify the problems or sections of the ecological system that we wish it to. But it will also exist in outline to be completed further, as required, in other sections of the system. Since our main immediate interest is growth, that of course is what we should expect to reveal in most detail and scrutinize most completely in the present context.

The other case studies mentioned could be analysed in similar fashions. There is no need to go over these fully—a few points will suffice. Both these investigations deal—like the Lake Dobson experiment—with factors that limit or augment production of organisms used by fish for food. These factors therefore influence fish growth. But, even more than in Lake Dobson, the environments of sea lochs and brackish water fish ponds are dynamic environments. Their temperatures, salinities and innumerable other physico-chemical characteristics are subject to great change. So studies of the production of organisms within them are difficult. Again, however, if certain facts can be pinned down, then certain inferences and conjectures become possible. The sea loch experiments showed that fertilization could, with fair certainty, increase plankton abundance and, with considerable certainty, benthic density and biomass. But the effect of fertilization on fish growth was not unequivocally demonstrated. Nevertheless, this is merely a methodological fault. The problem could have been evaluated more satisfactorily had the experimental design been better.

Another very useful discipline imposed by the study of systems composed of sub-systems is the avoidance of the mere collection of miscellaneous data for their own sake. A sense of the "stategy" of ecosystems compels the continual construction and reconstruction of schemata, the asking of pertinent questions, the posing of feasible hypotheses and the devising of appropriate models, around which data may be collected and assembled. Attention to a gradually evolving ground plan of a problem leads to caution and economy in the compiling of information (see Watt, 1966, 1968).

Apart from these few case studies there is an enormous amount of raw material on the subject of fish growth to which the methods of systems analysis could be applied. In particular, there are the related studies in Table 9.1; plus the fish cultural treatises by Hickling (1961, 1962), Schaeperclaus (1933) and Swingle (1947, 1952 etc.) already mentioned.

10 An Operational Programme for the Study of Fish Growth

I. DERIVATION OF THE PROGRAMME

A. The Individual and the System

It was suggested in Chapters 1 and 9 that the methods of systems analysis offered an approach appropriate to the investigation of numerous complicated ecological problems, a proposition that may now be reconsidered.

Recently, Watt (1966, 1968) examined the meaning of the term "system" and suggested—as Van Dyne (1966) had already done—that for the ecologist the appropriate biotic complex to consider as a system is the ecosystem. Watt particularized his example of an ecosystem as a forest with its contained biota. More importantly, he generalized the ecologist's idea of a system into the following form: "An interlocking complex of processes characterized by many reciprocal cause-effect pathways." Watt pointed out that the analysis of complex systems in which interacting variables play a prominent part has long been called for in such manifestly different fields as naval logistics, the construction of missiles, the operation of airlines and in the investigation of the physiology of respiratory and cardiovascular systems. He demonstrated that ideas central to systems analysis are also relevant to the analysis of complex ecological situations (ecosystems). Some examples follow:

(i) It is more advantageous to analyse a complex system into many simple functional "units" than into fewer relatively complex ones.

(ii) The "loops" known to computer programmers are based on recurrence relations, "in which the output for each stage in the computation is the input for the following stage"; this situation strikingly resembles that in many ecological systems.

(iii) Applications of principles of optimality and simulation of systems or parts of systems through models or schemes are among fundamentals of systems analysis; they are also methods obviously relevant in the examination of ecosystems.

(iv) Complex feedback control systems, primitively derived in engineering from steam-engine governors, thermostats and other simple servomechanisms, have been shown to be present in organisms, during the last century of progress in physiology: in recent years ecologists have revealed similar control systems in various population processes as well.

Watt (1966, 1968) has listed the steps in a systems research programme suitable for analysis of ecological situations: measurement, analysis, description, simulation and optimization. The frequently confusing complexity of so many ecological systems can make it extremely difficult to pick out for measurement those variables which are likely to be of most importance in the problem's interpretation. For example, it might be important to establish the timing of a process, with respect to a species' life cycle, before deciding on the importance of measuring its rate, intensity or duration. It is also important to minimize bias in measurement and obtain high statistical efficiency; here modern methods of instrumentation for automatic data collection play a most valuable part.

As far as analysis is concerned, multiple regressions and multiple analysis of variance lead naturally to the use of computers. This tendency is reinforced when dealing with ecological processes that recur, regularly or irregularly.

Description of systems really amounts to the construction of schemes, or models, in attempts to depict the operations of the systems or of the processes within the systems. A model may, as circumstances dictate, be mathematical and formal, mechanical or hydraulic, electronic, a plan on paper or a hybrid of these types. In simulation and optimization, a model's potential for accurately describing an ecosystem can be tested under a selected range of conditions. Destruction or irreparable change would frequently follow interference with an actual system. The model's usefulness may, therefore, sometimes be difficult to test against the system it is simulating. Occasionally a series of predictions about a system, based on a model, may have to be tested by keeping the system itself under critical surveillance for a long period—possibly many years. The proof of the validity or unsuitability of the model as a simulation of the system under such circumstances may require the passage of much time. But if the model seems to be successful it will be obvious not only that it has potential predictive value, which will facilitate the generation of new and better models, but it will also suggest that the original measurements and assumptions about the system were accurate. There is no reason, of course, why a number of models cannot be used simultaneously and in parallel to simulate a system. A selection process will then determine the "best" of these, indicate their common features or points of difference and show the direction for elaboration and improvement.

In this book, the dissection of growth processes in the individual fish has been stressed. And, using ideas based on the study of individual animals, an attempt has been made to show how these can be applied to growth as a population process. The main point, which underlies all proposals to use models in the solution of population growth problems, is that these can only be basically understood in terms of growth in individuals.

B. Data Acquisition

Approaching an understanding of ecological systems through model building and systems simulation implies massive collection of data. There is nothing new about this in any branch of biology but the methods of data collection which have become available in recent years are beginning to revolutionize the scope of ecology. For example, Kavanau (1966) has described apparatus for comprehensive monitoring of the activities of small mammals, by means of which quantified data may be automatically and simultaneously collected on variables such as distance run, time outside nest, water drunk, food consumed, excretion etc. By varying his array of sensors and recorders an investigator may construct a highly detailed sequential account of the combined activity regimes of the confined animals in relation to a preselected and programmed environment and, by varying the conditions of the environment, be able to observe and automatically record its effects in terms of changed activity. Perhaps even more significant is the possibility of the continuous recording of various aspects of activity over very extended periods. Thus, whereas early work on activity of small mammals was frequently marred or vitiated because the animals had insufficient time to become used to living in the activity chamber, present opportunities for continuous long-term monitoring frequently permit the ready identification of early atypical behaviour.

It is not implied that earlier ecologists did not make extended observations of relevant activity where possible. Park (1935) recorded diurnal activity in forest animals by the use of simple apparatus which allowed days or even weeks of activity data to be collected. Spoor (1946) used a metal vane mounted on a work-hardened wire for recording goldfish activity in an aquarium, through water currents generated in swimming. What was usually missing from the earlier work was the means for simultaneously sensing and recording various aspects of behaviour. The rise of electronic techniques and remote multichannel recording by pen and oscilloscope, combined with the use of sophisticated data storage and processing and computer analysis, are what have so greatly increased the scope of this approach. It is not that there has been a revolution in strategy or, essentially, even in technique but in the logistics of data collection and processing.

Savage (1966) has recently described the range of set-ups and apparatus for automatic data collection on animals and indicated the appropriate contexts for their employment in ecological research.

C. Construction of Models

If a major step in systems research is to devise a model then we should give a little general consideration to problems of modelling and of what may be hoped for by ecologists who resort to this technique.

Holling (1966) discussed model building strategy, taking several cases of predation as working examples. As a first step he compared several known predation patterns and from these constructed models to illustrate different sorts of "attack threshold" characterizing predators as diverse in type as mantids, sticklebacks and cats. These he compared with a theoretical model of a more generalized predator. Holling then constructed a flow diagram of predation for computer use. He pointed out that digital computers are eminently suited to the evaluation of such discontinuous, integral events as births, deaths, migration, predation, parasitism etc. Watt (1968) gives the general steps followed in construction of flow diagrams, together with examples.

In biology, a frequent shortcoming of models and schemes has been that "they combine a mixture of objects, processes and relations without systems or logic . . . as though they were logically . . . equivalent to components of the system represented." Kesteven and Ingpen (1966a, b) have strongly criticized this situation, using an "erroneous" diagram from a well-known ecological textbook which is supposed to illustrate "certain inanimate and animate influences involved in the metabolism of a lake community." The example is certainly typical of innumerable similar schemes in the ecological literature. Kesteven and Ingpen (1966a, b) have put forward a preliminary "language" for diagrammatic representation of biotic systems which offers symbols for systems as a whole, control systems within these, supply channels (for matter and energy) and identifier signs (arrows) to indicate direction or sense of actions. They have warned that:

> One should not represent . . . time relations on a diagram representing trophic, tropic or teleonomic relations. No attempt should be made to represent size in such diagrams; quantities—of material transfer, intensities of influences, etc.—should be presented as numbers or symbols placed beside relational lines.

Kesteven and Ingpen (1966a) also criticized the view sometimes expressed that models should, in their ultimate refinement or elaboration, be identical with the systems they are simulating. In terms of such reasoning a perfect model of an organism would be the organism. They observe that mathematical formulations of growth in populations or growth in laboratory experiments are both essentially models. Models may be inadequate because they lack certain components of the actual system they simulate or because of unreal values or relations assigned to factors or components of the system as represented. Models are tautologies; models should indicate trends, processes or mechanisms and not become in themselves a major concern of the investigator. Models should not become an increasingly elaborate analogue of the simulated biotic system, but should make it possible to bridge the gap between existing knowledge of the system's structure and internal relations and "a knowledge of the necessary outcome of the system under specified conditions."

Berman (1963) outlined the most acute uses of models, emphasizing their limitations. He pointed out (Berman, 1963a) that we cannot determine the "order" of a model—the particular equations or functions it should encompass and the number of its parameters—merely from considering the data available on a system. This follows because any given model can be held to represent a degenerate instance of a model of a higher order. The constraints placed on the choice of a model usually stem naturally from arriving at a minimal order of complexity for satisfactory prediction based on the type and complexity of data. Berman (1963) also sketched the use of a perturbation method for constructing mathematical models of systems. It may be assumed, especially where information is inadequate, that in disturbing a supposed system its various components behave according to the operation of a "minimum principle". This term signifies that a perturbation affects a minimum of the total number of parameters of the system or of "functionally and statistically independent variables ... related to the original parameters of the model." In whatever way a model is finally designed to simulate a system it may be tested in relation to further acquired data until a modification that most clearly satisfies the character of the disturbed system is derived (Berman, 1963).

Garfinkel (1962); Garfinkel and Sack (1964); Garfinkel, MacArthur and Sack (1964) have given interesting theoretical treatments of model simulation of simple ecological systems. Unfortunately these systems themselves were theoretical constructs rather than actual plant-animal complexes. Recently Watt (1968) has given a much more detailed and developed presentation of the methods of collection and organization of data for use in computer simulation of biological systems. Watt's work is of great value to ecologists in supplying for the first time examples of the application of simulation models and computer programmes to actual problems in ecological resource management, together with the mathematical background.

Figure 10.1 is an attempt to represent in a simple diagram some major factors influencing fish growth and some ways in which growth influences, in turn, the populations in which fish live. The diagram gives a superficial summary of various relationships held to exist between growth and other variables, which were discussed at length in preceding chapters. It also epitomizes certain axioms about these relationships.

The central place in the chart is occupied by growth, not of the population, but of the individual fish. The factors that directly affect its growth, such as supply of available food, efficiency of assimilation and the possible governing effects of temperature, are indicated.

A division of the chart by a major interface between the population and the individual is straddled by the factor of intensity of competition for food, because although competition is a population process (Chapter 5) it manifests its effects through the growth of the individual. Population age or size structures are labile properties which are also partly determined by growth of the individual. Egg productivity by adults is determined by age and size structures of population, and egg production is therefore governed

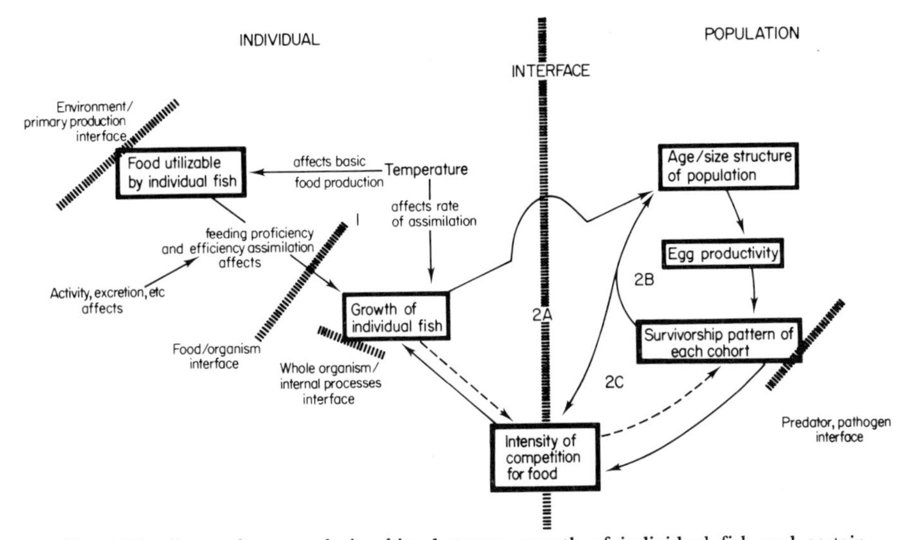

Fig. 10.1. Chart of some relationships between growth of individual fish and certain population parameters. Some environmental factors (some of them due to the presence of other fish) that affect fish growth are also shown. Interfaces are shown between processes that are functions of individuals and of populations, between the individual, its food and internal processes and between survivorship (a population statistic) and predators, pathogens etc. Area 1 indicates the group of processes dominated by temperature—all are functions of individual fish. 2A represents a cycle of processes in which both the growth of individuals affect or may be affected by population dynamics. 2B shows a loop which signifies that survivorship affects age/size structure and that, through population egg productivity, the pattern of survival of each cohort may be influenced. 2C is a cycle in which the principle of internal governance of population occurs through the process of competition. Where a broken arrow parallels a full one, action in the reverse sense to that signified by the latter is indicated and the relationship is also less certain. See text for additional discussion. (Original.)

partly by growth (Chapter 6). (Of course, the number of eggs laid will also depend on location, availability and abundance of suitable spawning substrates.) The survivorship pattern of each cohort (which may be deemed as starting its life as a certain number of fertile eggs) determines its fraction of the age/size structure of the population and eventually determines its own particular contribution to the adult breeding stock.

We have seen (Chapter 5) that intensity of competition for food is related to the biomass of the cohorts comprising the population and it is therefore obvious that both the survivorship pattern, which determines the number of a cohort alive at any given time, and the size (not the age) structure will establish the value of C_i for any given cohort using a particular food supply. In considering competition as expressed in terms of its effects on growth we inevitably recross the interface between population and individual.

A chart of the type of Fig. 10.1 is employable in two major ways. In the first place, a particular point of interest in a population—for instance, egg productivity—can be studied. Now Chapter 6 examined the role of growth, egg production and maintenance of populations. There is no need to reiterate the main points, but obviously the principal features of these relationships are implicit in the loop designated 2B in Fig. 10.1. This loop could be studied by itself, assuming a certain internal coherence and completeness. Indeed, this has frequently been a predominating interest in fish population dynamics (Cushing, 1968). If, however, deeper insight into the nature of these relationships and their mechanisms is required sections 2A and 2C of Fig. 10.1 must become involved and we begin to employ the scheme in the second major way.

Remember that intensity of competition for food occupies a position of unique importance in that it is affected by the numbers, sizes and (possibly) ages of fish in a population, while growth is in turn strongly affected by competition. Growth is a property of individual fish, so that we are therefore taken back and forth across the population/individual interface by virtue of this position of competition in the scheme.

As we tried to demonstrate in Chapters 2 and 4, growth of individuals is open not only to mathematical characterization but, more importantly, to quantitative empirical investigation. Such analysis, an example of applied physiology, involves an accounting of the various competing metabolic needs of the activities and functionings of the organism. The analysis can be carried to whatever depth of detailed information is required—at least in principle.

Consider, for example, section 1 of the chart. Chapters 2 and 4 traced the steps needed for understanding the fate of food potentially available to fish for growth. Figures 2.8, 2.9, 4.9 and Tables 2.2 to 2.4 showed how the food for standard metabolism, activity, growth, excretion etc. must all be found (from the "net energy" of the food ingested). The effects of temperature were also considered and, as Fig. 10.1 indicates, these may be mediated through augmented food production, modified activity (either that of the food search or extraneous) or by setting the actual rate of assimilation.

As earlier chapters showed, the physiological analysis of growth can be perfomed to whatever level is dictated by the particular type of ecological insight sought. It can halt at partition calorimetry measured directly or indirectly by respiratory exchange or by feeding experiments. We can investigate the role of endocrines in directing, synchronizing and imposing a particular tempo on growth. Problems of relative growth of different parts of the body may be important. Growth processes can also be viewed at cellular or sub-cellular levels if desired (Chapters 2 and 4) and it is clear that much analysis of these aspects of fish growth is urgently required.

In analysing growth *within* the individual we cross another interface; that between the overt growth of the whole organism and that fully demonstrable only through recourse to its internal processes.

There is also an interface between the fish and its food supply. To comprehend the reasons for the growth pattern in the individual, we must turn to the processes governing food production, which are separated by an interface between food and the environment in which it is produced, and examine the factors that influence the abundance of food. Chapter 9 was devoted to some consideration of this problem and no more need be added here.

Returning to the population side of the major interface of Fig. 10.1, the presence of an interface between the population and its predators and pathogens is noted. Chapter 7 was devoted to an examination of some of the effects different sorts of predation could have on population structure and the question was also discussed in Chapter 6.

So far, production has not been mentioned in connection with Fig. 10.1. Production by, or of, a fish population is, as should be apparent from earlier chapters, a result of growth and population dynamics. Figure 10.1 could, with little difficulty, be recast so as to feature production in the fish population in relation to growth and most of the other population and environmental parameters mentioned.

II. CONCLUDING REMARKS

Of the growing number of ecological problems which are benefitting from the attention of systems analysts (see, especially, Watt, 1966, 1968), only a few have so far been concerned with fish or fisheries. These have been reviewed by Watt who has also fully catalogued the research needs in this area, which include methods for preparing biomathematical models of ecological situations and their use in the form of computer languages so that large-scale computer simulations may be programmed.

The mobility of fish and the difficulties of catching, or sometimes even seeing, them detract from the study of their ecology. There is, therefore, a chronic need for large-scale data collection on a routine—and, if possible, automated—basis. Abundance or census counts of fish passing a fixed point in a river or the entry or exit to a lake, at a time of mass migration, have become routine. Modern technology is helping to make such data collections automatic and self-recording (Netboy, 1968).

It is perhaps pertinent to point out that, in the Northern Hemisphere, fishery data have been recorded abundantly for so long and so much is known, relatively speaking, about fisheries biology, that the value of massive data collection is sometimes glossed over. In countries like Australia, however, where there are quite inadequate fisheries records, vexing recurrent problems of fishery management often remain unresolved or require inordinately long periods for even their partial solution (Weatherley, 1971).

The relatively long lifetimes of fish, the great distances and spaces they frequently range over, the various sources of their food, enemies and competitors are inescapable facts. They combine to produce an urgent need for models and computer simulation programmes to aid in analysing the effects of these factors on population processes. The problems of growth will continue to play an important central role in these processes.

The schemes and considerations I have tried to advance in this book must not, in any sense, be regarded as attempts at proposing ecological "laws". They are simply "field maps" of a group of interlocked biological problems of inordinate complexity. They can be read at any level of detail dictated by particular needs of analysis or computation. The chapters of this book furnish some of the methodologic raw material to be employed in the study of the "block" or "black box" diagrams or charts such as Fig. 10.1, at whatever level of magnification appears appropriate to the situation.

References

Allee, W. C., Park, O., Emerson, A. E., Park, T. and Schmidt, K. P. (1949). "Principles of Animal Ecology." Saunders, Philadelphia.

Allen, K. R. (1935). The food and migration of the perch (*Perca fluviatilis*) in Windermere. *J. Anim. Ecol. 4,* 264-273.

Allen, K. R. (1938). Some observations on the biology of the trout (*Salmo trutta*) in Windermere. *J. Anim. Ecol. 7,* 333-349.

Allen, K. R. (1940). Studies of the biology of the early stages of the salmon (*Salmo salar*). 1. Growth in the River Eden. *J. Anim. Ecol 9,* 1-23.

Allen, K. R. (1941). Studies on the biology of the early stages of the salmon (*Salmo salar*). 2. *J. Anim. Ecol. 10,* 273-295.

Allen, K. R. (1950). The computation of production in fish populations. *N.Z. Sci. Rev. 8,* 89.

Allen, K. R. (1951). The Horokiwi Stream. A study of a trout population. *Fish. Bull. N.Z. No. 10.*

Allen, K. R. (1962). The natural regulation of population in the Salmonidae. *N.Z. Sci. Rev. 20,* 58-62.

Alm, G. (1946). Reasons for the occurrence of stunted fish populations with special regard to the perch. *Meddn St. Unders.–o. FörsAnst. SötvattFisk. 25,* 1-146.

Andrewartha, H. G. (1961). "Introduction to the Study of Animal Populations." Methuen, London.

Andrewartha, H. G. and Birch, L. C. (1954). "The Distribution and Abundance of Animals." Chicago University Press, Chicago.

Backiel, T. and Le Cren, E. D. (1967). Some density relationships for fish population parameters. *In* "The Biological Basis of Freshwater Fish Production" (S. D. Gerking, ed.). Blackwell, Oxford.

Bagenal, T. B. (1967). A short review of fish fecundity. *In* "The Biological Basis of Freshwater Fish Production" (S. D. Gerking, ed.). Blackwell, Oxford.

Bagenal, T. B. (1969). The relationship between food supply and fecundity in brown trout *Salmo trutta* L. *J. Fish Biol. 1,* 167-182.

Bainbridge, R. (1958). The speed of swimming of fish as related to size and to the frequency and amplitude of tail beat. *J. exp. Biol. 35,* 109-133.

Bainbridge, R. (1961). Problems of fish locomotion. *In* "Vertebrate Locomotion." (J. E. Harris, ed.). *Symp. zool. Soc. Lond. No. 5.*

Ball, J. N. (1969). Prolactin (fish prolactin or paralactin) and growth hormone. *In* "Fish Physiology. II. The Endocrine System" (W. S. Hoar and D. J. Randall, eds). Academic Press, London and New York.

Ball, R. C. (1948). Fertilization of natural lakes in Michigan. *Trans. Am. Fish. Soc. 78*, 145-155.

Bardach, J. (1955). Certain biological effects of thermocline shifts. *Hydrobiologia 7*, 309-324.

Barrington, E. J. W. (1957). The alimentary canal and digestion. *In* "The Physiology of Fishes" (M. E. Brown, ed.), Vol. 1. Academic Press, New York and London.

Bartell, R. J. (1968). The role of pheromones in the life cycle of light brown apple moth. *Proc. ecol. Soc. Aust. 3*, 155-158.

Barton Browne, L. (1968). The problem of obtaining physiological information for studies of insect ecology. *Proc. ecol. Soc. Aust. 3*, 159.

Beamish, F. W. H. (1964). Respiration of fishes with special emphasis on standard oxygen consumption. III. Influence of oxygen. *Can. J. Zool. 42*, 355-366.

Beamish, F. W. H. and Mookherjii, P. S. (1964). Respiration of fishes with special emphasis on standard oxygen consumption. I. Influence of weight and temperature on respiration of goldfish, *Carassius auratus* L. *Can. J. Zool. 42*, 161-175.

Beamish, F. W. H. and Dickie, L. M. (1967). Metabolism and biological production in fish. *In* "The Biological Basis of Freshwater Fish Production" (S. D. Gerking, ed.). Blackwell, Oxford.

Beckman, W. C. (1940). Increased growth of rock bass, *Ambloplites rupestris* (Rafinesque), following reduction in the density of the population. *Trans. Am. Fish. Soc. 70*, 143-148.

Beckman, W. C. (1942). Further studies on the increased growth rate of the rock bass *Ambloplites rupestris* (Rafinesque), following the reduction in density of the population. *Trans. Am. Fish. Soc. 72*, 72-78.

Beckman, W. C. (1946). The rate of growth and sex ratio for seven Michigan fishes. *Trans. Am. Fish. Soc. 76*, 63-81.

Beckman, W. C. (1948). Changes in growth rate of fishes following reduction in population densities by winter-kill. *Trans. Am. Fish. Soc. 78*, 82-90.

Berman, M. (1963). A postulate to aid in model building. *J. theor. Biol. 4*, 229-236.

Bertalanffy, L. von (1938). A quantitative theory of organic growth. *Hum. Biol. 10*, 181-213.

Bertalanffy, L. von (1951). An outline of general system theory. *Br. J. Phil. Sci 1*, 134-172.

Bertalanffy, L. von (1957). Quantitative laws in metabolism and growth. *Q. Rev. Biol. 32*, 217-231.

Bertalanffy, L. von (1960). Principles and theory of growth. *In* "Fundamental Aspects of Normal and Malignant Growth" (W. W. Nowinski ed.). Elsevier, Holland.

Beverton, R. J. H. and Holt, S. J. (1957). On the dynamics of exploited fish populations *Fishery Invest., Lond. Ser. II. Vol. XX.*

Beyers, R. J. and Odum, H. T. (1959). The use of carbon dioxide to construct pH curves for the measurement of productivity. *Limnol. Oceanogr 4*, 499-502.

Birch, L. C. (1957). The meanings of competition. *Am. Nat. 91*, 5-18.
Birch, L. C. and Ehrlich, P. R. (1967). Evolutionary history and population biology. *Nature, Lond. 214*, 349-352.
Blackburn, M. (1950). A biological study of the anchovy, *Engraulis australis* (White), in Australian waters. *Aust. J. mar. Freshwat. Res. 1*, 3-84.
Blair, W. F. (1964). The case for ecology. *BioScience 14*, 17-19.
Blažka, P., Volf, M. and Čepela, M. (1960). A new type of respirometer for the determination of the metabolism of fish in an active state. *Physiologia Bohemoslov. 9*, 553-558.
Bodenheimer, F. S. (1958). "Animal Ecology Today." Jünk, Netherlands.
Bonner J. T. (1965). "Size and Cycle: An Essay on the Structure of Biology." Princeton University Press, Princeton.
Borutsky, E. V. (1939a). Dynamics of the biomass of *Chironomus plumosus* in the profundal of Lake Beloie. *Trudў limnol. Sta. Kosine 22*, 156-195.
Borutsky, E. V. (1939b). Dynamics of the total benthic biomass in the profundal of Lake Beloie. *Trudў limnol. Sta. Kosine 22*, 196-218.
Braum, E. (1967). The survival of fish larvae in reference to their feeding behaviour and the food supply. *In* "The Biological Basis of Freshwater Fish Production" (S. D. Gerking, ed.). Blackwell, Oxford.
Brett, J. R. (1963). The energy required for swimming by young sockeye salmon with a comparison of the drag force on a dead fish. *Trans. R. Soc. Can. 1* (Ser. IV), 441-457.
Brett, J. R. (1964). The respiratory metabolism and swimming performance of young sockeye salmon. *J. Fish. Res. Bd Can. 21*, 1183-1226.
Brett, J. R. (1965). The swimming energetics of salmon. *Scient. Am. 213*, 80-87.
Brett, J. R., Shelbourn, J. E. and Shoop, C. T. (1969). Growth rate and body composition of fingerling sockeye salmon, *Oncorhynchus nerka*, in relation to temperature and ration size. *J. Fish. Res. Bd Can. 26*, 2363-2394.
Brody, S. (1927). Growth and development. III. Growth rates, their evolution and significance. *Bull. Mo. agric. Exp. Stn. 105*, 1-70.
Brody, S. (1945). "Bioenergetics and Growth." Reinhold, New York.
Brook, A. J. and Holden, A. V. (1957). Fertilization experiments in Scottish freshwater lochs. I. Loch Kinardochy. *Freshwat. Salm. Fish. Res. No. 17.*
Brown, G. D. (1968). The nitrogen and energy requirements of the euro (*Macropus robustus*) and other species of macropod marsupials. *Proc. ecol. Soc. Aust. 3*, 106-112.
Brown, M. E. (1946a). The growth of brown trout (*Salmo trutta* Linn.). I. Factors influencing the growth of trout fry. *J. exp. Biol. 21*, 118-129.
Brown, M. E. (1946b). The growth of brown trout (*Salmo trutta* Linn.). II. The growth of two-year-old trout at a constant temperature of 11·5°C. *J. exp. Biol. 21*, 130-142.
Brown, M. E. (1946c). The growth of brown trout (*Salmo trutta* Linn.). III. The effect of temperature on the growth of two-year old trout. *J. exp. Biol. 21*, 145-155.
Brown, M. E. (1957). Experimental studies on growth. *In* "The Physiology of Fishes" (M. E. Brown, ed.), Vol. 1. Academic Press, New York and London.
Bullough, W. S. (1971). Ageing of mammals. *Nature, Lond. 229*, 608-610.
Bullough, W. S. and Laurence, E. B. (1967). Epigenetic mitotic control. *In* "Control of Cellular Growth in Adult Organisms" (H. Teir and T. Rytömaa, eds). Academic Press, London and New York.

Carlander, K. D. (1955). The standing crop of fish in lakes. *J. Fish. Res. Bd Can. 12*, 543-570.

Carlander, K. D. and Cleary, R. E. (1949). The daily activity patterns of some freshwater fish. *Am. Midl. Nat. 41*, 447-452.

Chapman, C. B. and Mitchell, J. H. (1965). The physiology of exercise. *Scient. Am. 212* (5), 88-96.

Chapman, D. W. (1967). Production in fish populations. *In* "The Biological Basis of Freshwater Fish Production" (S. D. Gerking, ed.). Blackwell, Oxford.

Chapman, D. W. (1968). Production. *In* "Methods for Assessment of Fish Production in Fresh Waters" (W. E. Ricker, ed.). IBP Handbook No. 3. Blackwell, Oxford.

Cheek, D. B. (1968a). Cellular growth: Introduction. *In* "Human Growth: Body Composition, Cell Growth, Energy and Intelligence" (D. B. Cheek, ed.). Lea and Febiger, Philadelphia.

Cheek, D. B. (1968b). Muscle cell growth in abnormal children. *In* "Human Growth; Body Composition, Cell Growth, Energy and Intelligence" (D. B. Cheek, ed.). Lea and Febiger, Philadelphia.

Cheek, D. B. (1968c). Muscle cell growth in normal children. *In* "Human Growth: Body Composition, Cell Growth, Energy and Intelligence" (D. B. Cheek, ed.). Lea and Febiger, Philadelphia.

Cheek, D. B., Brasel, J. A. and Graystone, J. E. (1968). Muscle cell growth in rodents: sex differences and the role of hormones. *In* "Human Growth: Body Composition, Cell Growth, Energy and Intelligence" (D. B. Cheek, ed.). Lea and Febiger, Philadelphia.

Chen, F. Y. (1965). The living-space effect and its economic implications. Appendix III, in *Rep. trop. Fish Cult. Res. Inst. Malacca*.

Chen, F. Y. and Prowse, G. A. (1966). The effect of living space on the growth rate of fish. *Ichthyologica. 3*, 11-20.

Clements, F. E. and Shelford, V. H. (1939). "Bio-Ecology." Wiley, New York.

Cocking, A. W. (1957). Relation between the upper ultimate lethal temperature and the temperature range for good health in the roach (*Rutilis rutilis*). *Nature, Lond. 100*, 661-662.

Cocking A. W. (1959a). The effects of high temperatures on roach (*Rutilis rutilis*). I. The effects of constant high temperature. *J. exp. Biol. 36*, 203-216.

Cocking, A. W. (1959b). The effects of high temperatures on roach (*Rutilis rutilis*). II. The effects of temperature increasing at a known constant rate. *J. exp. Biol. 36*, 217-226.

Cole, L. C. (1954). The population consequences of life history phenomena. *Q. Rev. Biol. 29*, 103-137.

Creac'h, Y. (1966). Protein thiols and free amino acids of carp tissues during prolonged starvation. *Archs Sci physiol. 20*, 115-121.

Creac'h, Y. and Cournede, C. (1965). Contribution to the study of enforced starvation in the carp, *Cyprinus carpio* L.; variations in the amount of water and nitrogen in the tissues. *Bull. Soc. Hist. nat. Toulouse. 100*, 361-370.

Cureton, T. (1951). "The Physical Fitness of Champion Athletes." Illinois University Press Urbana.

Curtis, H. J. (1963). Biological mechanisms underlying the aging process. *Science, New York 141*, 686-694.

Cushing, D. H. (1968). "Fisheries Biology: A Study in Population Dynamics." University Wisconsin Press.

Das, S. M. and Moitra, S. K. (1956a). Studies on the food of some common fishes of Uttar Pradesh, India. Part II. *Proc. natn. Acad. Sci. India 26,* 213-223.

Das, S. M. and Moitra, S. K. (1956b). Studies on the food of some common fishes of the Uttar Pradesh, India. Part III. *Proc. natn. Acad. Sci. India 26,* 224-233.

Davies, P. M. C. (1963). Food input and energy extraction efficiency in *Carassius auratus. Nature, Lond. 198,* 707.

Davies, P. M. C. (1964). The energy relations of *Carassius auratus* L. I. Food input and energy extraction efficiency at two experimental temperatures. *Comp. Biochem. Physiol. 12,* 67-79.

Davies, P. M. C. (1966). The energy relations of *Carassius auratus* L. II. The effect of food, crowding and darkness on heat production. *Comp. Biochem. Physiol. 17,* 893-995.

Davies, P. M. C. (1967). The energy relations of *Carassius auratus* L. III. Growth and the overall balance of energy. *Comp. Biochem. Physiol. 23,* 59-64.

Dawes, B. (1930-31a). Growth and maintenance in the plaice (*Pleuronectes platessa* L.). Part I. *J. mar. biol. Ass. U.K. 17,* 103-174.

Dawes, B. (1930-31b). Growth and maintenance in the plaice (*Pleuronectes platessa* L.). Part II. *J. mar. biol. Ass. U.K. 17,* 877-975.

DeBach, P. (1966). The competitive displacement and coexistence principles. *A. Rev. Ent. 11,* 183-212.

Deelder, C. L. (1951). A contribution to the knowledge of the stunted growth of perch (*Perca fluviatilis* L.) in Holland. *Hydrobiologia 3,* 357-378.

Deevey, E. S. (1947). Life tables for natural populations of animals. *Q. Rev. Biol. 22,* 283-314.

Egglishaw, H. J. (1970). Production of salmon and trout in a stream in Scotland. *J. Fish Biol. 2,* 117-136.

Elton, C. S. (1927). "Animal Ecology." Sidgwick and Jackson, London.

Elton, C. S. (1966). "The Pattern of Animal Communities." Methuen, London.

El-Zarka, Salah el-Din (1959). Fluctuations in the population of yellow perch, *Perca flavescens* (Mitchell), in Saginaw Bay Lake Huron, *U.S. Dept. Int. Fish. Bull. 151.*

Engelmann, M. D. (1966). Energetics, terrestrial field studies and animal productivity. *In* "Advances in Ecological Research" (J. B. Cragg, ed.), Vol. 3. Academic Press, London and New York.

Fairbridge, W. S. (1951). The New South Wales tiger flathead *Neoplatycephalus macrodon* (Ogiliby). I. Biology and age determination. *Aust. J. mar. Freshwat. Res. 2,* 117-178.

Farmer, G. J. and Beamish, F. W. H. (1969). Oxygen consumption of *Tilapia nilotica* in relation to swimming speed and salinity. *J. Fish. Res. Bd Can. 26,* 2807-2821.

Fell, H. B. (1962). Some effects of environment on epidermal differentiation. *J. Derm. 74,* 1.

Fell, H. B. and Mellanby, E. (1953). Metaplasia produced in cultures of chick ectoderm by high Vitamin A. *J. Physiol., Lond. 119,* 470-488.

Fell, H. B. and Rinaldi, L. N. (1965). The effects of vitamins A and C on cells and tissues in culture. *In* "Cells and Tissues in Culture. I. Methods, Biology and Physiology" (E. N. Wilmer, ed.). Academic Press, London and New York.

Foerster, R. E. (1944). The relation of lake population density to size of young sockeye salmon (*Oncorhynchus nerka*). *J. Fish Res. Bd Can. 6,* 267-280.

Fortunatova, K. R. (1961). Effect of predacious fish upon the structure of commercial fish populations. Trudy Sovesch. po dinamike chislennosti ryb. Akad, Nauk SSSR.

Frost, W. E. (1946). On the food relationships of fish in Windermere. *Biol. Jaarb. 13,* 216-231.

Frost, W. E. (1954). The food of the pike *Esox lucius* L. in Windermere. *J. Anim. Ecol. 23,* 339-360.

Frost, W. E. and Brown, M. E. (1967). "The Trout." Collins, London.

Frost W. E. and Kipling, C. (1959). The determination of the age and growth of pike (*Esox lucius* L.) from scale and opercular bones. *J. Cons. perm. int. Explor. Mer. 24,* 314-341.

Frost, W. E. and Smyly, W. J. P. (1952). The brown trout of a moorland fishpond. *J. Anim. Ecol. 21,* 62-68.

Fry, F. E. J. (1957). The aquatic respiration in fish. *In* Part I of "The Physiology of Fishes" (M. E. Brown, ed.). Academic Press, New York and London.

Fry, F. E. J. and Hart, J. S. (1948). Cruising speed of goldfish in relation to water temperature. *J. Fish. Res. Bd Can. 7,* 169-175.

Fryer, G. (1959). The trophic interrelationships and ecology of some littoral communities of Lake Nyasa with special reference to the fishes, and a discussion of the evolution of a group of rock-frequenting Cichlidae. *Proc. zool. Soc. Lond. 132,* 153-281.

Gagianut, B. (1951). Investigations on the influence of steroids on growth. Effects of hormones of the adrenal cortex. *Schweiz. Z. allg. Path. Bakt. 14,* 66.

Gammon, J. R. and Hasler, A. D. (1965). Predation by introduced muskellunge on perch and bass, I: years 1-5. *Wis. Acad. Sci. Arts Let. 54,* 249-272.

Garfinkel, D. (1962). Digital computer simulation of ecological systems. *Nature, Lond. 194,* 856-857.

Garfinkel, D., MacArthur, R. H. and Sack, R. (1964). Computer simulation and analysis of simple ecological systems. *Ann. N.Y. Acad. Sci. 115,* 943-951.

Garfinkel, D. and Sack, R. (1964). Digital computer simulation of an ecological system based on a modified mass action law. *Ecology 45,* 502-507.

Gauld, D. T. (1950). A fish cultivation experiment in an arm of a sea-loch. III. The plankton of Kyle Scotnish. *Proc. R. Soc. Edinb. LXIV,* 36-64.

Gause, G. F. (1934). "The Struggle for Existence." Williams and Wilkins, Baltimore.

Geng, H. (1925). Der Futterwert de natürlichen Fishnahrung Ztschr. *J. Fisch. 23,* 137-165.

Gerking, S. D. (1952). The protein metabolism of sunfishes of different ages. *Physiol. Zoöl. 25,* 137-165.

Gerking, S. D. (1954). The food turnover of a bluegill population. *Ecology 35,* 490-498.

Gerking, S. D. (1955a). Influence of rate of feeding on body composition and protein metabolism of bluegill sunfish. *Physiol. Zoöl. 28,* 267-282.

Gerking, S. D. (1955b). Endogenous nitrogen excretion of bluegill sunfish. *Physiol. Zoöl. 28,* 283-289.

Gerking, S. D. (1962). Production and food utilization in a population of bluegill sunfish. *Ecol. Monogr, 32,* 31-78.

Gerking, S. D. (1966). Annual growth cycle, growth potential and growth compensation in the bluegill sunfish in northern Indiana lakes. *J. Fish. Res. Bd Can. 23*, 1923-1956.

Gerking, S. D. (ed.) (1967). "The Biological Basis of Freshwater Fish Production." Blackwell, Oxford.

Gibson, M. B. and Hirst B. (1955). The effect of salinity and temperature on the pre-adult growth of guppies. *Copeia 3*, 241-243.

Gorbman, A. (1969). Thyroid function and its control in fishes. *In* "Fish Physiology. II. The Endocrine System" (W. S. Hoar and D. J. Randall, eds). Academic Press, London and New York.

Goss, R. J. (1964). "Adaptive Growth." Academic Press, New York and London.

Goss, R. J. (1967). The strategy of growth. *In* "Control of Cellular Growth in Adult Organisms" (H. Teir and T. Rytomaa, eds). Academic Press, London and New York.

Gould, B. S. (1961). Ascorbic acid-independent and ascorbic acid-dependent collagen-forming mechanisms. *Ann. N.Y. Acad. Sci. 92*, 168-174.

Graham, M. (1929). Studies of age-determination in fish. Part II. A survey of the literature. *Fishery Invest., Lond. Ser. II. Vol. II 2*, 1-11.

Graham, M. (ed.) (1956). "Sea Fisheries—Their Investigation in the United Kingdom." Edward Arnold, London.

Gray, J. (1957). How fishes swim. *Scient. Am. 197*, 48-54.

Gray, J. (1936). Studies in animal locomotion. VI. The propulsive powers of the dolphin. *J. exp. Biol. 13*, 192-199.

Greenwood, P. H. (1963). "A History of Fishes," 2nd edition (original edition by J. R. Norman). Benn, London.

Greer-Walker, M. (1970). Growth and development of the skeletal muscle fibres of the cod (*Gadus morhua* L.) *J. Cons. perm. int. Explor. Mer. 33*, 228-244.

Gross, F. (1947). An experiment in marine fish cultivation: V. Fish growth in a fertilised sea loch (Loch Craiglin). *Proc. R. Soc. Edinb. LXIII*, 56-95.

Gross, F. (1950). A fish cultivation experiment in an arm of a sea-loch. V. Fish growth in Kyle Scotnish. *Proc. R. Soc. Edinb. LXIV*, 109-135.

Hartley, P. H. T. (1948). Food and feeding relationships in a community of freshwater fishes. *J. Anim. Ecol. 33*, 1-14.

Hasler, A. D. and Bardach, J. E. (1949). Daily migrations of perch in Lake Mendota, Wisconsin. *J. Wildl. Mgmt. 13*, 40-51.

Hayflick, L. (1966). Senescence and cultured cells. *In* "Perspectives in Experimental Gerontology" (N. W. Shock, ed.). Thomas, Springfield.

Hayflick, L. (1968). Human cells and ageing. *Scient. Am. 218*, 32-37.

Henderson, H. F., Hasler, A. D. and Chipman, G. G. (1966). An ultrasonic transmitter for use in studies of movements of fish. *Trans. Am. Fish. Soc. 95*, 350-356.

Hergenrader, G. L. and Hasler, A. D. (1967). Seasonal changes in swimming rates of yellow perch in Lake Mendota as measured by sonar. *Trans. Am. Fish. Soc. 96*, 373-383.

Herrington, W. C. (1948). Limiting factors for fish populations. Some theories and an example. *In* "A Symposium on Fish Populations." *Bull. Bingham oceanogr. Coll. 11*, 229-283.

Hickling, C. F. (1961). "Tropical Inland Fisheries." Trop. Agr. Ser. Longmans, London.

Hickling, C. F. (1962). "Fish Culture." Publishers Assn., Blackmore Press, London.

Hile, R. (1936). Age and growth of the cisco, *Leucichthys artedi* (Le Seur), in the lakes of the north-eastern highlands, Wisconsin. *Bull. Bur. Fish., Wash. 48*, 211-317.

Hodgson, W. C. (1957). "The Herring and its Fishery." Routledge and Kegan Paul, London.

Holling, C. S. (1966). The strategy of building models of complex ecological systems. *In* "Systems Analysis in Ecology" (K. E. F. Watt, ed.). Academic Press, New York and London.

Holt, S. J. (1962). The application of comparative population studies to fishery biology—an exploration. *In* "The Exploitation of Natural Animal Populations" (E. D. Le Cren and M. W. Holdgate, eds). Blackwell, Oxford.

Hughes, R. D. (1968). Ecologists and physiology. *Proc. ecol. Soc. Aust. 3*, 139-146.

Hughes, R. D. and Walker, D. (1965). Education and training in ecology. *Vestes 8*, 173-178.

Huntsman, A. G. (1948). Fishing and assessing populations. Article 1 in "A symposium on fish populations". *Bull. Bingham oceanogr. Coll. 11*, 5-25.

Hutchinson, G. E. (1957). Concluding remarks. *Symp. quant. Biol. XXII.*

Hutchinson, G. E. (1965). "The Ecological Theater and the Evolutionary Play." Yale University Press.

Huxley, J. S. (1932). "Problems of Relative Growth." Methuen, London.

Idler D. R. and Clemens, W. A. (1959). The energy expenditure of Fraser River salmon during the spawning migration to Chilko and Stuart Lakes. *Prog. Rep. int. Pacif. Salm. Fish. Commn. 1959.*

Ivlev, V. S. (1939a). The energy balance of the growing larva of *Siluris glonis Dokl. (Proc.) Acad. Sci. U.S.S.R. 25*, 87-89.

Ivlev, V. S. (1939b). The effect of starvation on energy transformation during the growth of fish. *Dokl. (Proc.) Acad. Sci. U.S.S.R. 25*, 90-92.

Ivlev, V. S. (1945). The biological productivity of waters (Transl. W. E. Ricker). *Usp. sovrem. Biol. 19*, 98-120.

Ivlev, V. S. (1961). "Experimental Ecology of the Feeding of fishes" (Transl. D. Scott). Yale University Press, New Haven.

Ivlev, V. S. (1964). *In* "Techniques for the Investigation of Fish Physiology" E. N. Pavlovskii, ed.). Transl. from Russian by Israel Program Sci. Transls. U.S. Dept. Int. and Nat. Sci. Found., Washington.

Johnson, W. R. (ed.) (1960). "Science and Medicine of Exercise and Sports." Harper, N.Y.

Jones, R. (1958). Lee's phenomenon of "apparent change in growth-rate" with particular reference to haddock and plaice. *In* "Some Problems for Biological Fishery Survey and Techniques for their Solution," pp. 229-242. Int. Comm. N. Atlantic Fish. Spec. Publ. No. 1.

Juday, C., Schloemer, C. L. and Livingstone, C. (1938). Effect of fertilizers on plankton production and on fish growth in a Wisconsin lake. *Progr. Fish Cult. 40*, 24-27.

Karlson, P. (1963). "Introduction to Biochemistry." Academic Press, New York and London.

Karpovich, P. (1965). "Physiology of Muscular Activity." Saunders, Philadelphia.

Karzinkin, G. S. and Krivobok, M. N. (1964). Balance sheet experiments on nitrogen metabolism of fish. *In* "Techniques for the Investigation of Fish

Physiology" (E. N. Pavlovskii, ed.). Acad. Sci. U.S.S.R. Ichthyol. Bd. Translated by Israel Program for Sci. Transl. U.S. Dept. Commerce, Washington.

Kavanau, J. L. (1966). Automatic monitoring of the activities of small mammals. *In* "Systems Analysis in Ecology" (K. E. F. Watt, ed.). Academic Press, New York and London.

Keast, A. and Webb, D. (1966). Mouth and body form relative to feeding ecology in the fish fauna of a small lake, Lake Opinicon, Ontario. *J. Fish. Res. Bd Can. 23,* 1845-1874.

Kesteven, G. L. and Ingpen, R. R. (1966a). The representation of relations, including those of sociality, in biotic systems. *Proc. Ecol. Soc. Aust. 1,* 79-83.

Kesteven, G. L. and Ingpen, R. R. (1966b). Representation of the structure of biotic systems. *Aust. J. Sci. 29,* 97-102.

Kinne, O. (1960). Growth, food intake, and food conversion in a euryplastic fish exposed to different temperatures and salinities. *Physiol. Zoöl. 33,* 288-317.

Kipling, C. and Frost, W. E. (1970). A study of the mortality, population numbers, year class strengths, production and food consumption of pike, *Esox lucius* L., in Windermere from 1944 to 1962. *J. Anim. Ecol. 39,* 115-157.

Kitamura, S., Suwa, T., Ohara, S. and Nakagawa, K. (1967). Studies on vitamin requirements of rainbow trout—II. The deficiency symptoms of fourteen kinds of vitamin. *Bull. Jap. Soc. scient. Fish. 33,* 1120-1125.

Kleiber, M. (1961). "The Fire of Life—an Introduction to Animal Energetics." Wiley, New York.

Knobil, E. and Hotchkiss, J. (1964). Growth and hormones. *A. Rev. Physiol. 26,* 47-74.

Kovalevskaya, L. A. (1956). "Energetika Dvizhushcheisya Ryby" (The energy of fish in motion). Sbornik posvyashchennyi pamyati akad. P. P. Lazareva, Izdatel'stvo AN SSSR.

Kutty, M. N. (1968). Respiratory quotients in goldfish and rainbow trout. *J. Fish. Res. Bd Can. 25,* 1689-1728.

Lack, D. (1954). "The Natural Regulation of Animal Numbers." Oxford University Press, Oxford.

Lack, D. (1966). "Population Studies of Birds." Oxford University Press, Oxford.

Lack, D. (1968). "Ecological Adaptation for Breeding in Birds." Methuen, London.

Lagler, K. F. (1968). Capture, sampling and examination of fishes. *In* "Methods for Assessment of Fish Production in Fresh Waters" (W. E. Ricker, ed.). I.B.P. Handbook No. 3. Blackwell, Oxford.

Lake, J. S. (1957). Trout populations and habitats in New South Wales. *Aust. J. mar. Freshwat. Res. 8,* 414-450.

Langford, R. R. (1948). Fertilization of lakes in Algonquin Park, Ontario. *Trans. Am. Fish. Soc. 78,* 133-144.

Lasnitzki, I. (1965). The action of hormones on cell and organ cultures. *In* "Cells and Tissues in Culture. I. Methods, Biology and Physiology" (E. N. Willmer ed.). Academic Press, London and New York.

LeBrasseur, R. J. (1969). Growth of juvenile chum salmon (*Oncorhynchus keta*) under different feeding regimes. *J. Fish. Res. Bd Can. 26,* 1631-1645.

Lea, E. (1910). On the methods used in the herring investigations. *Publ. Circonst. Cons. perm. int. Explor. Mer. 53,* 7-25.

Lea, E. (1913). Further studies concerning the method of calculating the growth of herrings. *Publ. Circonst. Cons. perm. int. Explor. Mer. 66.*

Lea, E. (1938). A modification of the formula for the growth of herring. *Rapp. P.-v. Réun. Cons. perm. int. Explor. Mer. 63.*

Le Cren, E. D. (1947). The determination of the age and growth of the perch (*Perca fluviatilis*) from the opercular bone. *J. Anim. Ecol. 16,* 188-204.

Le Cren, E. D. (1951). The length-weight relationship and seasonal cycle in gonad weight and condition in the perch (*Perca fluviatilis*). *J. Anim. Ecol. 20,* 201-219.

Le Cren, E. D. (1958). Observations on the growth of perch (*Perca fluviatilis* L.) over twenty-two years, with special reference to the effects of temperature and changes in population density. *J. Anim. Ecol. 27,* 287-334.

Le Cren, E. D. (1962). The efficiency of reproduction and recruitment in freshwater fish. *In* "The Exploitation of Natural Animal Populations" (E. D. Le Cren and M. W. Holdgate, eds). Blackwell, Oxford.

Le Cren, E. D. (1965). Some factors regulating the size of populations of freshwater fish. *Mitt. Verein. theor. angew. Limnol. 13,* 88-105.

Lee R. M. (1912). An investigation into the methods of growth determination in fishes. *Publ. Circonst. Cons. perm. int. Explor. Mer. 63.*

Lee, R. M. (1920). A review of the methods of age and growth determination in fishes by means of scales. *Fishery Invest., Lond. 4.*

Lindeman, R. L. (1942). The trophic-dynamic aspect of ecology. *Ecology 23,* 339-418.

Lister, D. and McCance, R. A. (1967). Severe undernutrition in growing and adult animals. 17. The ultimate results of rehabilitation: pigs. *Br. J. Nutr. 21,* 787-799.

Lister, D., Cowan, T. and McCance, R. A. (1966). Severe undernutrition in growing and adult animals. 16. The ultimate results of rehabilitation: poultry. *Br. J. Nutr. 20,* 633-648.

Love, R. M. (1957). The biochemical composition of fish. *In* "The Physiology of Fishes—Vol. 1" (M. E. Brown, ed.). Academic Press, New York and London.

Love, R. M. (1970). "The Chemical Biology of Fishes." Academic Press, London and New York.

Macan, T. T. (1949). Survey of a moorland fishpond. *J. Anim. Ecol. 18,* 160-186.

McCance, R. A. and Widdowson, E. M. (1962). Nutrition and growth. *Proc. R. Soc. 156,* 326-337.

McFadden, J. T. (1961). A population study of the brook trout, *Salvelinus fontinalis. Wildl. Monogr., Chestertown 7.*

McFadden, J. T. and Cooper, E. L. (1964). Population dynamics of brown trout in different environments. *Physiol. Zoöl. 37,* 355-363.

Macfadyen, A. (1963). "Animal Ecology: Aims and Methods." Pitman, London.

Magnuson, J. J. (1962). An analysis of aggressive behaviour, growth, and competition for food and space in medaka (*Oryzias latipes* (Pisces, Cyprinodontidae)). *Can. J. Zool. 40,* 313-363.

Main, A. R. (1968). Physiology in the management of kangaroos and wallabies. *Proc. ecol. Soc. Aust. 3,* 96-105.

Maitland, P. S. (1965). The feeding relationships of salmon, trout, minnows, stone loach and three-spined sticklebacks in the River Endrick, Scotland. *J. Anim. Ecol. 34,* 109-133.

Mann, K. H. (1967). The cropping of the food supply. *In* "The Biological Basis

of Freshwater Fish Production" (S. D. Gerking, ed.). Blackwell, Oxford.
Mann, R. H. K. and Orr, D. R. O. (1969). A preliminary study of the feeding relationships of fish in a hard-water and a soft-water stream in southern England. *J. Fish. Biol. 1,* 31-44.
Marshall, A. J. (ed.) (1960-61). "Biology and Comparative Physiology of Birds." 2 Vols. Academic Press, New York and London.
Marshall, N. B. (1965). "The Life of Fishes." Weidenfeld and Nicholson, London.
Marshall, S. M. (1947). An experiment in marine fish cultivation: III. The plankton of a fertilized loch. *Proc. R. Soc. Edinb. LXIII,* 21-33.
Medawar, P. B. (1945). Size, shape and age. *In* "Essays on Growth and Form Presented to D'Arcy Wentworth Thompson" (W. E. Le Gros Clark and P. B. Medawar, eds). Oxford University Press.
Miller, R. S. (1967). Pattern and process in competition. *Adv. Ecol. Res. 4,* 1-74.
Milne, A. (1961). Definition of competition among animals. *In* "Mechanisms in Biological Competition." Symposia Soc. Exp. Biol. XV.
Mitchell, H. H. (1962-64). "Comparative Nutrition of Man and Domestic Animals." 2 Vols. Academic Press, New York and London.
Molnar, G. (1967). The gastric digestion of living, predatory fish. *In* "The Biological Basis of Freshwater Fish Production" (S. D. Gerking, ed.). Blackwell, Oxford.
Morehouse, L. E. and Miller, A. T. (1967). "Physiology of Exercise," 5th ed. Mosby, St. Louis.
Mortimer, C. H. and Hickling, C. F. (1954). Fertilizers in fishponds. *Fishery Publs. colon. Off. 5.*
Moscona, A. A. (1965). Recombination of dissociated cells and the development of cell aggregates. *In* "Cells and Tissues in Culture" (E. N. Willmer, ed.), Vol. 1. Academic Press, London and New York.
Munro, W. (1957). The pike of Loch Choin. *Freshwat. Salm. Fish. Res. 16.*
Myers, K. (1967). Ecology and behaviour—a review of the ways in which Australian ecologists use behaviour in their studies. *Proc. ecol. Soc. Aust. 2,* 79-98.
Myers, K. (1968). Physiology and rabbit ecology. *Proc. ecol. Soc. Aust. 3,* 1-7.
Needham, A. E. (1964). "The Growth Processes in Animals." Pitman, London.
Neill, R. M. (1938). The food and feeding of the brown trout (*Salmo trutta* L.) in relation to the organic environment. *Trans. R. Soc. Edinb. 59,* 481-520.
Nelson, P. R. and Edmondson, W. T. (1955). Limnological effects of fertilizing Bare Lake, Alaska. *Fishery Bull. Fish Wildl. Serv. U.S. 102.*
Netboy, A. (1968). "The Atlantic Salmon—A Vanishing Species?" Faber and Faber, London.
Nicholls, A. G. (1957). The Tasmanian trout fishery. I. Sources of information and treatment of data. *Aust. J. mar. Freshwat. Res. 8,* 451-475.
Nicholls, A. G. (1958a). The Tasmanian trout fishery. II. The fishery of the north-west region. *Aust. J. mar. Freshwat. Res. 9,* 19-59.
Nicholls, A. G. (1958b). The Tasmanian trout fishery. III. The rivers of the north and east. *Aust. J. mar. Freshwat. Res. 9,* 167-190.
Nicholls, A. G. (1958c). The egg yield from brown and rainbow trout in Tasmania. *Aust. J. mar. Freshwat. Res. 9,* 526-536.
Nicholls, A. G. (1961). The Tasmanian trout fishery. IV. The rivers of the south-east. *Aust. J. mar. Freshwat. Res. 12,* 17-53.

Nicholson, A. J. (1933). The balance of animal populations. *J. Anim. Ecol. 2*, 132-178.

Nicholson, A. J. (1954). An outline of the dynamics of animal populations. *Aust. J. Zool. 2*, 9-65.

Nicholson, A. J. (1957). The self-adjustment of populations to change. *In* "Population Studies: Animal Ecology and Demography." *Symp. quant. Biol. 22*, 153-173.

Nikolskii, G. V. (1969). "Theory of Fish Population Dynamics as the Biological Background for Rational Exploitation and Management of Fishery Resources." Oliver and Boyd, Edinburgh.

Nikolskii, G. V. (1963). "The Ecology of Fishes." Academic Press, London and New York.

Nutman, S. R. (1950). A fish cultivation experiment in an arm of a sea-loch. II. Observations on some hydrographic factors in Kyle Scotnish. *Proc. R. Soc. Edinb. LXIV*, 5-35.

Odum, E. P. (1959). "Fundamentals of Ecology." Saunders, Philadelphia.

Odum, E. P. (1964). The new ecology. *BioScience 14*, 14-16.

Odum, H. T. (1956). Primary production in flowing waters. *Limnol. oceanogr. 1*, 102-117.

Odum, H. T. (1957a). Trophic structure and productivity of Silver Springs, Florida. *Ecol. Monogr. 27*, 55-112.

Odum, H. T. (1957b). Primary production in eleven Florida springs and a marine turtle-grass community. *Limnol. Oceanogr. 2*, 85-97.

Odum, H. T. and Hoskin, C. M. (1957). Metabolism of a laboratory stream microcosm. *Instn. mar. Sci. IV*, 115-133.

Odum, H. T. and Hoskin, C. M. (1958). Comparative studies on the metabolism of marine waters. *Instn. mar. Sci. V*, 16-46.

Odum, H. T. and Odum, E. P. (1955). Trophic structure and productivity of a windward coral reef community on Eniwetock Atoll. *Ecol. Monogr. 25*, 291-320.

Olsen, A. M. (1954). The biology, migration, and growth rate of the school shark (*Galeorhinus australis* Macleay) (Carcharhanidae) in south-eastern Australian waters. *Aust. J. mar. Freshwat. Res. 5*, 353-410.

Orr, A. P. (1947). An experiment in marine fish cultivation: II. Some physical and chemical conditions in a fertilized sea-loch (Loch Craiglin, Argyll). *Proc. R. Soc. Edinb. LXIII*, 3-20.

Paget, G. W. (1920). Report on the scales of some teleostean fish, with special reference to their method of growth. *Fishery Invest., Lond. 4.*

Paloheimo, J. E. and Dickie, L. M. (1965). Food and growth of fishes. I. A growth curve derived from experimental data. *J. Fish. Res. Bd Can. 22*, 521-542.

Paloheimo, J. E. and Dickie, L. M. (1966a). Food and growth of fishes. II. Effects of food and temperature on the relation between metabolism and body weight. *J. Fish. Res. Bd Can. 23*, 869-908.

Paloheimo, J. E. and Dickie, L. M. (1966b). Food and growth of fishes. III. Relations among food, body size, and growth efficiency. *J. Fish. Res. Bd Can. 23*, 1209-1248.

Pandian, T. J. (1967). Intake, digestion, absorption and conversion of food in the fishes *Megalops cyprinoides* and *Ophiocephalus striatus. Mar. Biol. 1*, 16-32.

Park, O. (1935). Studies in nocturnal ecology. III. Recording apparatus and further analysis of activity rhythm. *Ecology 16*, 152-163.

Parrish, B. P. (1958). Convenor's report—comparison of European and North American techniques of measuring nets, of reading ages of fish and of studying growth. *In* "Some Problems for Biological Fisheries Survey and Techniques for their Solution." Plus articles by members of the working Party on methods in fishery research. Spec. Publ. No. 1. Int. Commn. Northwest. Atlantic Fish.

Patriarche, M. E. (1968). Production and theoretical equilibrium yields for the bluegill (*Lepomis macrochirus*) in two Michigan lakes. *Trans. Am. Fish. Soc. 99*, 242-251.

Pearl, R. (1927). The growth of populations. *Q. Rev. Biol. 2*, 532-548.

Pentelow, F. T. K. (1939). The relation between growth and food consumption in the brown trout (*Salmo trutta*). *J. exp. Biol. 16*, 446-473.

Pickford, G. E. and Atz, J. W. (1957). "The Physiology of the Pituitary Gland of Fishes." N.Y. Zool. Soc., New York.

Popova, O. A. (1967). The "predator-prey" relationship among fish. *In* "The Biological Basis of Freshwater Fish Production" (S. D. Gerking, ed.). Blackwell, Oxford.

Rao, G. M. M. (1968). Oxygen consumption of rainbow trout (*Salmo gairdneri*) in relation to activity and salinity. *Can. J. Zool. 46*, 781-785.

Raymont, J. E. G. (1947). An experiment in marine fish cultivation: IV. The bottom fauna and the food of flatfishes in a fertilized sea loch (Loch Craiglin). *Proc. R. Soc. Edinb. LXIII*, 34-55.

Raymont, J. E. G. (1950). A fish cultivation experiment in an arm of a sea-loch. IV. The bottom fauna of Kyle Scotnish. *Proc. R. Soc. Edinb. LXIV*, 65-108.

Ricker, W. E. (1946). Production and utilization of fish populations. *Ecol. Monogr. 16*, 373-391.

Ricker, W. E. (1948). "Methods of Estimating Vital Statistics of Fish Populations." Ind. Univ. Publ. Sci. Ser. No. 15.

Ricker, W. E. (1954). Stock and recruitment. *J. Fish. Res. Bd Can. 11*, 559-623.

Ricker, W. E. (1958). "Handbook of Computations for Biological Statistics of Fish Populations." *Bull. Fish. Res. Bd Can. 119.*

Ricker, W. E. (1963). Big effects from small causes: two examples from fish population dynamics. *J. Fish. Res. Bd Can. 20*, 257-264.

Ricker, W. E. (ed.) (1968). "Methods for Assessment of Fish Production in Fresh Waters." IBP Handbook No. 3. Blackwell, London.

Ricker, W. E. and Foerster, R. E. (1948). Computation of fish production. *Bull. Bingham oceanogr. Coll. 11*, 173-221.

Robertson, W. van B. (1961). The biological role of ascorbic acid in connective tissue. *Ann. N.Y. Acad. Sci. 92*, 159-167.

Rose, E. T. and Moen, T. (1952). The increase in game-fish populations in East Okoboji Lake, Iowa, following intensive removal of rough fish. *Trans. Am. Fish. Soc. 82*, 104-114.

Ruskin, B., Pomerat, C. N. and Roskin, A. (1951). Toxicity of various cortisone preparations on embryonic chick heart, spleen and spinal cord in tissue culture. *Tex. Rep. Biol. Med. 9*, 786-795.

Savage, J. C. (1966). Telemetry and automatic data acquisition systems. *In* "Systems Analysis in Ecology" (K. E. F. Watt, ed.). Academic Press, New York and London.

Savage, M. (1961). "The Biology and Life History of the Common Frog." Pitman, London.

Schaeperclaus, W. (1933). "Textbook of Pond Culture." Paul Parey, Berlin (Transl. from the German by F. Hund and Issued as *Fishery Leafl. Fish Wildl. Serv. U.S. 311*).

Schmitz, W. R. and Hetfeld, R. E. (1965). Predation by introduced muskellunge on perch and bass, II: years 8-9. *Wis. Acad. Arts Sci. Let. 54*, 273-282.

Schuster, W. H. (1952). "Fish Culture in Brackish-Water Ponds of Java." Indo. Pac. Fish. Counc. Spec. Publ. No. 1.

Sears, P. B. (1964). Ecology—a subservsive subject. *BioScience 14*,11-13.

Shapovalov, L. and Taft, A. C. (1954). The life histories of the steelhead rainbow trout (*Salmo gairdneri gairdneri*) and Silver salmon (*Oncorhynchus kisutch*) with special reference to Waddell Creek, California, and recommendations regarding their management. *Fish Bull. Calif. 98*.

Shelford, V. H. (1913). "Animal Communities in Temperate America, as illustrated in the Chicago Region; a study in Animal Ecology." Chicago University Press, Chicago.

Shetter, D. S. and Hazzard, A. S. (1939). Species composition by age-groups and stability of fish populations in sections of three Michigan trout streams during the summer of 1937. *Trans. Am. Fish. Soc. 68*, 281-302.

Shock, N. W. (1962). The physiology of aging. *Scient. Am. 206*, 100-110.

Shorygin, A. A. (1946). Seasonal dynamics of food competition of fishes. *Zool. Zh. Ukr. 25*, 441-450.

Slobodkin, L. (1962). "Growth and Regulation of Animal Populations." Holt, Rinehart and Winston, New York.

Smit, H. (1965). Some experiments on the oxygen consumption of goldfish (*Carassius auratus* L.) in relation to swimming speed. *Can. J. Zool. 43*, 623-633.

Smith, M. W. (1945). Preliminary observations upon the fertilization of Crecy Lake, New Brunswick. *Trans. Am. Fish. Soc. 75*, 165-174.

Smith, M. W. (1948). Fertilization of a lake to improve trout angling. *Atlant. Biol. Stn. Note 105*.

Smith, M. W. (1955). Fertilization and predator control to improve trout angling in natural lakes. *J. Fish. Res. Bd Can. 12*, 210-237.

Smith, S. H. (1968). Species succession and fishery exploitation in the Great Lakes. *J. Fish. Res. Bd Can. 25*, 667-693.

Smyly, W. J. P. (1952). Observations on the food of the fry of perch (*Perca fluviatilis* Linn.) in Windermere. *Proc. zool. Soc. Lond. 122*, 407-416.

Spoor, W. A. (1946). A quantitative study of the relationship between the activity and oxygen consumption of the goldfish, and its application to the measurement of respiratory metabolism in fishes. *Biol. Bull. mar. biol. Lab., Woods Hole 91*, 312-325.

Stevens, E. D. and Fry, F. E. J. (1970). The rate of thermal exchange in a teleost, *Tilapia mossambica. Can. J. Zool. 48*, 221-226.

Stott, B. (1968). Marking and tagging. *In* "Methods for Assessment of Fish Production in Fresh Waters" (W. E. Ricker, ed.). IBP Handbook No. 3. Blackwell, London.

Swift, D. R. (1955). Seasonal variations in the growth rate, thyroid gland activity and food reserves of brown trout (*Salmo trutta* Linn.). *J. exp. Biol. 32*, 751-764.

Swift, D. R. (1961). The annual growth-rate cycle in brown trout (*Salmo trutta* Linn.). *J. exp. Biol. 38*, 591-604.
Swift, D. R. (1962). Evidence for the absence of an endogenous growth rate rhythm in brown trout (*Salmo trutta* Linn.). *Comp. Biochem. Physiol. 6*, 91-93.
Swift, D. R. (1964). The effect of temperature and oxygen on the growth rate of the Windermere char (*Salvelinus alpinus willughbii*). *Comp. Biochem. Physiol. 12*, 179-183.
Swift, D. R. Pickford, G. E. (1965). Seasonal variations in the hormone content of the pituitary gland of the perch, *Perca fluviatilis* L. *Gen. Comp. Endocrinol. 5*, 345-365.
Swingle, H. S. (1946). Experiments with combinations of largemouth black bass, bluegills, and minnows in ponds. *Trans. Am. Fish. Soc. 76*, 46-62.
Swingle, H. S. (1947). Experiments on pond fertilization. *Bull. Ala. agric. Exp. Stn. 264.*
Swingle, H. S. (1949). Some recent developments in pond management. *Trans N. Am. Wildl. Conf. 14*, 295-312.
Swingle, H. S. (1950). Relationships and dynamics of balanced and unbalanced fish populations. *Bull. Ala. agric. Exp. Stn. 274.*
Swingle, H. S. (1951). Experiments with various rates of stocking bluegills, *Lepomis macrochirus* Rafinesque, and largemouth bass, *Micropterus salmoides* (Lacépède), in ponds. *Trans. Am. Fish. Soc. 80*, 218-230.
Swingle, H. S. (1952). Farm pond investigations in Alabama. *J. Wildl. Mgmt. 16*, 243-249.
Swingle, H. S. and Smith, E. V. (1939). Fertilizers for increasing the natural food for fish in ponds. *Trans. Am. Fish. Soc. 68*, 216-235.
Swingle, H. S. and Smith, E. V. (1940). Experiments on the stocking of fish ponds. *Trans. N. Am. Wildl. Conf. 15*, 267-276.
Swingle, H. S. and Smith, E. V. (1942). Management of farm fish ponds. *Bull. Ala. agric. Exp. Stn. 254.*
Swingle, H. S. and Smith, E. V. (1950). Factors affecting the reproduction of bluegill bream and largemouth black bass in ponds. *Ala. Circ. agric. Exp. Stn. 87.*
Swynnerton, G. H. and Worthington, E. B. (1939). Brown-trout growth in the Lake District. *Salm. Trout Mag. 97*, 337-355.
Tanner, J. M. (1955). "Growth at Adolescence." Blackwell, Oxford.
Tanner, J. M. (1960). "Human Growth." Vol. III. *In* "Symposia for the Study of Human Biology." Butterworths, London.
Teir, H. and Rytömaa, T. (eds) (1967). "Control of Cellular Growth in Adult Organisms." Academic Press, London and New York.
Teir, H., Lahtiharju, A., Alho, A. and Forsell, K. J. (1967). Autoregulation of growth by tissue breakdown products. *In* "Control of Cellular Growth in Adult Organisms" (H. Teir and T. Rytömaa, eds). Academic Press, London and New York.
Tesch, F. W. (1968). Age and growth. *In* "Methods for Assessment of Fish Production in Fresh Waters" (W. E. Ricker, ed.). IBP Handbook No. 3. Blackwell, London.
Thompson, D'Arcy W. (1942). "On Growth and Form," Vols. 1 and 2. Cambridge University Press, Cambridge.

Thompson, H. (1923). Problems in haddock biology with special reference to the validity and utilization of the scale theory. *Scient. Invest. Fishery Bd. Scotl. 1922, 5.*

Thompson, H. (1929). General features in the biology of the haddock (*G. aeglefinus*) in Icelandic waters in the period 1903-26. *Rapp. P.-V. Réun. Cons. perm int. Explor. Mer. 57*, 3-73.

Thompson, W. R. (1939). Biological control and the theories of the interactions of populations. *Parasitology 31*, 299-388.

Thomson, J. M. (1951). Growth and habits of the sea mullet, *Mugil dobula* Gunther, in Western Australia. *Aust. J. mar. Freshwat. Res. 2*, 193-225.

Thomson, J. M. (1957). Interpretation of the scales of the yellow-eye mullet, *Aldrichetta forsteri* (Cuvier and Valenciennes) Mugilidae. *Aust. J. mar. Freshwat. Res. 8*, 14-28.

Thomson, J. M. (1962). The tagging and marking of marine animals in Australia. *C.S.I.R.O. Div. Fish. Oceanogr. tech. Pap. 13.*

Ullyett, G. C. (1950). Competition for food and allied phenomena in sheep blowfly populations. *Phil. Trans. R. Soc. 234*, 77-174.

Ursin, E. (1967). A mathematical model of some aspects of fish growth, respiration and mortality. *J. Fish. Res. Bd Can. 24*, 2355-2453.

Van Dyne, G. M. (1966). Ecosystems, systems ecology, and systems ecologists. *Atom. Energy Biophys. Biol. Med. 3957*, 1-31.

Van Oosten, J. (1928). Life history of the lake herring (*Leucichthys artedi* Le Seuer) of Lake Huron as revealed by its scales, with a critique of the scale method. *Bull. Bur. Fish., Wash. 44*, 265-427.

Van Oosten, J. (1941). The age and growth of freshwater fishes. *In* "Symposium of Hydrobiology." Madison, Wisconsin.

Varley, M. E. (1967). "British Freshwater Fishes." Fishing News (Books), London.

Volterra, V. (1931). "Lecons sur la Theorie Mathematique de la Butte pour la Vie." Paris.

Wagner, A. F. and Folkers, K. (1964). "Vitamins and Coenzymes." Wiley, New York.

Walford, L. A. (1946). A new graphic method for describing the growth of animals. *Biol. Bull. mar. Biol. Lab., Woods Hole 91*, 312-325.

Warren, C. E. and Davis, G. E. (1967). Laboratory studies on the feeding, bioenergetics, and growth of fish. *In* "The Biological Basis of Freshwater Fish Production" (S. D. Gerking, ed.). Blackwell, Oxford.

Watt, K. E. F. (1955). Studies on population productivity. I. Three approaches to the optimum yield problem in populations of *Tribolium confusum. Ecol. Monogr. 25*, 269-290.

Watt, K. E. F. (1966). "Systems Analysis in Ecology." Academic Press, New York and London.

Watt, K. E. F. (1968). "Ecology and Resource Management." McGraw-Hill, New York.

Weatherley, A. H. (1958). Tasmanian farm dams in relation to fish culture. *C.S.I.R.O. Div. Fish. and Oceanogr. tech. Pap. 4.*

Weatherley, A. H. (1959). Some features of the biology of the tench *Tinca tinca* (Linnaeus) in Tasmania. *J. Anim. Ecol. 28*, 73-87.

Weatherley, A. H. (1963). Notions of niche and competition among animals, with special reference to freshwater fish. *Nature, Lond. 197*, 14-17.

Weatherley, A. H. (1965). Operation of a "law" of parsimony in shaping animal life cycles. *Nature, Lond. 207*, 804-806.

Weatherley, A. H. (1966). The ecology of fish growth. *Nature, Lond. 212*, 1321-1324.

Weatherley, A. H. (1971). Fish and fisheries. *In* "Conservation" (A. B. Costin and H. J. Frith, eds). Penguin, Australia.

Weatherley, A. H. and Lake, J. S. (1967). Introduced fish species in Australian inland waters. *In* "Australian Inland Waters and their Fauna: Eleven Studies" (A. H. Weatherley, ed.). A.N.U. Press, Canberra.

Weatherley, A. and Nicholls, A. G. (1955). The chemical enrichment of a lake. *Aust. J. mar. Freshwat. Res. 6*, 443-468.

White, A., Handler, P. and Smith, E. L. (1964). "Principles of Biochemistry." McGraw-Hill, Tokyo.

Widdowson, E. M. (1964). Early maturation and later development. *In* "Diet and Bodily Constitution" (G. E. W. Wolstenholme and M. O'Connor, eds). Ciba Foundtn. Study Conf. No. 17. Churchill, London.

Wilhelmi, A. E. (1955). Comparative biochemistry of growth hormone from ox, sheep, pig, horse, and fish pituitaries. *In* "The Hypophysial Growth Hormone, Nature and Actions" (R. W. Smith, O. H. Gaebler and G. N. H. Long, eds). McGraw-Hill, New York.

Williams, W. T. (1967). The computer botanist. *Aust. J. Sci. 29*, 266-271.

Winberg, G. G. (1956). Rate of metabolism and food requirements of fishes. *Fish. Res. Bd Can. Transl. Ser. 194.*

Windell, J. T. (1967). Rates of digestion in fishes. *In* "The Biological Basis of Freshwater Fish Production" (S. D. Gerking, ed.). Blackwell, Oxford.

Wohlschag, D. E. (1961). Respiratory metabolism and ecological characteristics of some fishes in McMurdo Sound, Antarctica. *Biol. Antarctic Res. Ser. 1*, 33-62.

Wolf, K. and Quimby, M. C. (1969). Fish cell and tissue culture. *In* "Fish Physiology. III. Reproduction and Growth, Bioluminescence, Pigments and Poisons" (W. S. Hoar and D. V. Randall, eds). Academic Press, London and New York.

Wuenscher, J. E. (1969). Niche specification and competition modelling. *J. theor. Biol. 25*, 436-443.

Wynne-Edwards, V. C. (1962). "Animal Dispersion in Relation to Social Behaviour." Oliver and Boyd, London.

Author Index

Numbers followed by an asterisk refer to the page on which the reference is listed.

Subject Index